T0190256

IFIP Advances in Information and Communication Technology

611

Editor-in-Chief

Kai Rannenberg, Goethe University Frankfurt, Germany

Editorial Board Members

TC 1 – Foundations of Computer Science
  Luís Soares Barbosa, *University of Minho, Braga, Portugal*

TC 2 – Software: Theory and Practice
  Michael Goedicke, University of Duisburg-Essen, Germany

TC 3 – Education
  Arthur Tatnall, *Victoria University, Melbourne, Australia*

TC 5 – Information Technology Applications
  Erich J. Neuhold, University of Vienna, Austria

TC 6 – Communication Systems
  Burkhard Stiller, University of Zurich, Zürich, Switzerland

TC 7 – System Modeling and Optimization
  Fredi Tröltzsch, TU Berlin, Germany

TC 8 – Information Systems
  Jan Pries-Heje, Roskilde University, Denmark

TC 9 – ICT and Society
  David Kreps, *National University of Ireland, Galway, Ireland*

TC 10 – Computer Systems Technology
  Ricardo Reis, *Federal University of Rio Grande do Sul, Porto Alegre, Brazil*

TC 11 – Security and Privacy Protection in Information Processing Systems
  Steven Furnell, *Plymouth University, UK*

TC 12 – Artificial Intelligence
  Eunika Mercier-Laurent, *University of Reims Champagne-Ardenne, Reims, France*

TC 13 – Human-Computer Interaction
  Marco Winckler, *University of Nice Sophia Antipolis, France*

TC 14 – Entertainment Computing
  Rainer Malaka, University of Bremen, Germany

IFIP – The International Federation for Information Processing

IFIP was founded in 1960 under the auspices of UNESCO, following the first World Computer Congress held in Paris the previous year. A federation for societies working in information processing, IFIP's aim is two-fold: to support information processing in the countries of its members and to encourage technology transfer to developing nations. As its mission statement clearly states:

IFIP is the global non-profit federation of societies of ICT professionals that aims at achieving a worldwide professional and socially responsible development and application of information and communication technologies.

IFIP is a non-profit-making organization, run almost solely by 2500 volunteers. It operates through a number of technical committees and working groups, which organize events and publications. IFIP's events range from large international open conferences to working conferences and local seminars.

The flagship event is the IFIP World Computer Congress, at which both invited and contributed papers are presented. Contributed papers are rigorously refereed and the rejection rate is high.

As with the Congress, participation in the open conferences is open to all and papers may be invited or submitted. Again, submitted papers are stringently refereed.

The working conferences are structured differently. They are usually run by a working group and attendance is generally smaller and occasionally by invitation only. Their purpose is to create an atmosphere conducive to innovation and development. Refereeing is also rigorous and papers are subjected to extensive group discussion.

Publications arising from IFIP events vary. The papers presented at the IFIP World Computer Congress and at open conferences are published as conference proceedings, while the results of the working conferences are often published as collections of selected and edited papers.

IFIP distinguishes three types of institutional membership: Country Representative Members, Members at Large, and Associate Members. The type of organization that can apply for membership is a wide variety and includes national or international societies of individual computer scientists/ICT professionals, associations or federations of such societies, government institutions/government related organizations, national or international research institutes or consortia, universities, academies of sciences, companies, national or international associations or federations of companies.

More information about this series at https://link.springer.com/bookseries/6102

Vallidevi Krishnamurthy ·
Suresh Jaganathan ·
Kanchana Rajaram ·
Saraswathi Shunmuganathan (Eds.)

Computational Intelligence in Data Science

4th IFIP TC 12 International Conference, ICCIDS 2021
Chennai, India, March 18–20, 2021
Revised Selected Papers

Springer

Editors
Vallidevi Krishnamurthy ⓘ
Sri Sivasubramaniya Nadar College of
Engineering
Chennai, India

Suresh Jaganathan ⓘ
Sri Sivasubramaniya Nadar College of
Engineering
Chennai, India

Kanchana Rajaram ⓘ
Sri Sivasubramaniya Nadar College of
Engineering
Chennai, India

Saraswathi Shunmuganathan ⓘ
Sri Sivasubramaniya Nadar College of
Engineering
Chennai, India

ISSN 1868-4238   ISSN 1868-422X (electronic)
IFIP Advances in Information and Communication Technology
ISBN 978-3-030-92602-1   ISBN 978-3-030-92600-7 (eBook)
https://doi.org/10.1007/978-3-030-92600-7

© IFIP International Federation for Information Processing 2021
This work is subject to copyright. All rights are reserved by the Publisher, whether the whole or part of the
material is concerned, specifically the rights of translation, reprinting, reuse of illustrations, recitation,
broadcasting, reproduction on microfilms or in any other physical way, and transmission or information
storage and retrieval, electronic adaptation, computer software, or by similar or dissimilar methodology now
known or hereafter developed.
The use of general descriptive names, registered names, trademarks, service marks, etc. in this publication
does not imply, even in the absence of a specific statement, that such names are exempt from the relevant
protective laws and regulations and therefore free for general use.
The publisher, the authors and the editors are safe to assume that the advice and information in this book are
believed to be true and accurate at the date of publication. Neither the publisher nor the authors or the editors
give a warranty, expressed or implied, with respect to the material contained herein or for any errors or
omissions that may have been made. The publisher remains neutral with regard to jurisdictional claims in
published maps and institutional affiliations.

This Springer imprint is published by the registered company Springer Nature Switzerland AG
The registered company address is: Gewerbestrasse 11, 6330 Cham, Switzerland

Preface

The main objective of the Fourth International Conference on Computational Intelligence in Data Science (ICCIDS), held virtually during March 18–20, 2021, was to help you connect the dots that matter to you. The world appears increasingly complex, and to seize the opportunities ahead we need to be better at looking at the world from a wider array of perspectives.

The growth of data, both in structured and unstructured form, presents challenges as well as opportunities for industry and academia over the next few years. With the explosive growth in data volume, it is essential that real-time information that is of use to businesses is extracted, in order to deliver better insights to decision-makers and to understand complex patterns.

God's greatest gift to humans is "intelligence". While human intelligence aims to adapt to new environments by utilizing a combination of different cognitive processes, artificial intelligence (AI) aims to build machines that can mimic human behavior and perform human-like actions. The human brain is not digital, but machines are. That is hard computing. On the other hand, computational intelligence (CI) is a new concept for advanced information processing. Computational intelligence tools offer adaptive mechanisms that enable the understanding of data in complex and changing environments. Hence, learning computational intelligence in data science helps researchers to gain knowledge and skills in order to analyze, interpret, and visualize huge volumes of data which are complex in nature.

The building blocks of computational intelligence involve computational modeling, natural intelligent systems, multi-agent systems, hybrid intelligent systems, etc.

Data is the key ingredient for the development and enhancement of all intelligence-based algorithms, and blockchain secures this data. Blockchain, IoT, and AI/CI can and should be applied jointly. One possible connection between these technologies could be that IoT collects and provides data, blockchain provides the infrastructure and sets up the rules of accessing the data, and computational intelligence optimizes the processes and rules.

The aim of ICCIDS 2021 was to explore computational intelligence and blockchain technology and how they can be used together. The conference program included a pre-conference workshop session along with prominent keynote talks and paper presentations.

The conference received a total of 75 submissions from authors all over India and a few papers from authors at universities outside India too, out of which 20 papers (an acceptance rate of 26.66%) were selected by the Program Committee after a careful and thorough review process. The papers were presented across two sessions. Two papers from each of the sessions were declared as the best papers from the respective sessions.

Sasikumar Venkatesh, Senior Software Engineer from Walmart Global Tech, India, elaborated on the topic "How Deep Learning Helps Finding Patterns in Time-Series

Data? – Exploring with Sparkling-Water (H2O.ai)" during the pre-conference workshop.

To enlighten our participants in this new era of blockchain enabled AI, five keynote talks were arranged. On the first day of the conference, Elizabeth Chang, Professor of Logistics and IT at the University of New South Wales, Australia, delivered her talk on "Enterprise Blockchain - Trust, Security and Privacy".

On the second day of the conference, a keynote talk on "Data Science for Social Good" was delivered by Ponnurangam Kumaraguru from IIIT Delhi, India. Another keynote talk on the topic of "Emerging Trends in Advanced Artificial Intelligence" was given by Vivek Singhal, Co-Founder and Chief Data Scientist of CellStrat, India.

On the third day of the conference, the program featured a keynote talk by Garrick Hileman, Head of Research at Blockchain.com, UK. He shared his knowledge about "Cryptocurrency and Blockchain Technology: Past, Present and Future". The final keynote was delivered by R.K. Agrawal, Professor at Jawaharlal Nehru University, India, on the topic "Deep Learning Models for Medical Image Analysis".

ICCIDS 2021 was organized by the Department of Computer Science and Engineering at Sri Sivasubramaniya Nadar College of Engineering (SSNCE), India; we are grateful to everyone who helped to make it a success, including all authors and participants.

March 2021   Kanchana R.
                                                       Suresh Jaganathan
                                                             Vallidevi K.
                                                           Saraswathi S.

Organization

Executive Committee

Chief Patron

Shiv Nadar   SSN Institutions, India

Patron

Kala Vijayakumar   SSN Institutions, India

General Chairs

V. E. Annamalai   SSNCE, India
Chitra Babu   SSNCE, India

Program Committee

Conference Chairs

K. Vallidevi   SSNCE, India
J. Suresh   SSNCE, India

Program Chairs

R. Kanchana   SSNCE, India
S. Saraswathi   SSNCE, India

Organizing Co-chairs

Y. V. Lokeswari   SSNCE, India
D. Venkatavara Prasad   SSNCE, India
G. Raghuraman   SSNCE, India
V. Balasubramanian   SSNCE, India

Workshop Speaker

Sasikumar Venkatesh   Walmart Global Tech, India

Session I Coordinator

N. Sujaudeen   SSNCE, India

Session I Chairs

T. T. Mirnalinee   SSNCE, India
D. Thenmozhi   SSNCE, India
B. Prabavathy   SSNCE, India

Session II Coordinator

R. Priyadharshini   SSNCE, India

Session II Chairs

A. Chamundeswari   SSNCE, India
V. S. Felix Enigo   SSNCE, India
S. Kavitha   SSNCE, India

Technical Program Committee

Ammar Mohammed	Cairo University, Egypt
Chua Chin Heng Matthew	National University of Singapore, Singapore
Sanjay Misra	Covenant University, Nigeria
Shamona Gracia Jacob	Nizwa College of Technology, Oman
Premkumar K.	IITDM Kancheepuram, India
Sivaselvan B.	IITDM Kancheepuram, India
Surendiran B.	NIT Puducherry, India
Umarani Jayaraman	IITDM Kancheepuram, India
Ramadoss B.	NIT Trichy, India
E. Uma	Anna University, India
Sriram Kailasam	IIT Mandi, India
Veena T.	NIT Goa, India
Subrahmanyam K.	K L University, India
Lakshmi C.	SRM Institute of Science and Technology, India
Latha Parthiban	Pondicherry University, India
Latha Karthika	Brandupwise Marketing, New Zealand
Arvind Ram A.	Google, USA
Sindhu Raghavan	Microsoft, USA
Venkatesh S.	Oracle, USA
Srinidhi Rajagopalan	InterSystems, USA
Venkatesh Sakamuri	Oracle, USA
Srikanth Bellamkonda	Oracle, USA
Sriraghav K.	Dell Inc, Chennai, India
Kamakshi Prasad V.	Jawaharlal Nehru Technological University, India
Anitha R.	Sri Venkateswara College of Engineering, India
Renukadevi Saravanan	VIT Chennai, India.
M. Anuradha	St. Joseph's College of Engineering, India
R. Anuradha	Sri Ramakrishna Engineering College, India
Deva Priya M.	Sri Krishna College of Technology, India
Gayathri K. S.	Sri Venkateswara College of Engineering, India
Menaka Pushpa	St. Joseph's Institute of Technology, India
Karthikeyan P.	Velammal College of Engineering and Technology, India
Sountharrajan S.	Bannari Amman Institute of Technology, India
Farida Begam Mohamed Hussain	CMR Institute of Technology, India

Contents

Machine Learning (ML), Deep Learning (DL), Internet of Things (IoT)

A Scalable Data Pipeline for Realtime Geofencing Using Apache Pulsar

K. Sundar Rajan, A. Vishal$^{(\boxtimes)}$, and Chitra Babu

Department of Computer Science and Engineering, Sri Sivasubramaniya Nadar
College of Engineering, Chennai, Tamil Nadu, India
{sundarrajan16110,vishal16124}@cse.ssn.edu.in, chitra@ssn.edu.in

Abstract. A geofence is a virtual perimeter for a real-world geographic area. Geofencing is a technique used to monitor a geographical area by dividing it into smaller subareas demarcated by geofences. It can be used to create triggers whenever a device moves across a geofence to provide useful location-based services. Since real-world objects tend to move continuously, it is essential to provide these services in real-time to be effective. Towards this objective, this paper introduces a scalable data pipeline for geofencing that can reliably handle and process data streams with high velocity using Apache Pulsar - an open-source Publish/Subscribe messaging system that has both stream processing and light-weight computational capabilities. Further, an implementation of the proposed data pipeline for a specific real-world case study is presented to demonstrate the envisaged advantages of the same.

Keywords: Geofencing · Apache pulsar · Stream processing · Scalable data pipeline

1 Introduction

Data science as a field has evolved rapidly over the years to solve increasingly complex problems that facilitate improved living standards for society. In recent times, data is being progressively generated in tremendous volume, velocity and variety. Analyzing this vast data can provide valuable business insights, which can lead to effective decision-making. There is a significant requirement for adequate computational resources to find quick answers to real-time queries involving big data.

Usually, such queries have been executed using sequential scans over a large fraction of a database. In the context of big data, this approach takes a lot of computation time. Increasingly, several applications demand real-time response rates. One example could be updating ads based on recent trends observed on Twitter and Facebook.

© IFIP International Federation for Information Processing 2021
Published by Springer Nature Switzerland AG 2021
V. Krishnamurthy et al. (Eds.): ICCIDS 2021, IFIP AICT 611, pp. 3–14, 2021.
https://doi.org/10.1007/978-3-030-92600-7_1

Stream handling and processing in real-time is a necessity in a lot of real-world applications. Real-time geofencing is one such example. Geofencing is a technique used to monitor a geographical area by dividing it into smaller parts demarcated by virtual boundaries known as geofences. The software which uses GPS, RFID, Wi-Fi, or cellular data can be used to trigger pre-programmed action such as - prompting mobile push notifications, triggering text messages or alerts, sending targeted ads on social media, allow tracking of fleets of vehicles, whenever a mobile device or RFID tag breaches a geofence set up around a geographical location.

Objects transmit their live location as a continuous data stream. This data needs to be processed sequentially and incrementally on a record-by-record basis. Message queues or publish/subscribe systems such as Apache Kafka [5], and RabbitMQ [15] are used to ingest the data stream in order to enable asynchronous communication as well as provide data persistence and fault-tolerance. This stream of location data should be transformed into a suitable format before analytics is performed to provide meaningful insights. Therefore, to perform such transformations in real-time, stream processing systems are such as Apache Spark [10] and Apache Flink [1] are used.

However, in certain use cases such as real-time geofencing, where a simple transformation from a specific geofenced area is required, such heavy computation frameworks incur excessive overhead. Therefore, Apache Pulsar [16] - an open-source Publish/Subscribe messaging system that has both stream processing and light-weight computational capabilities - is appropriate for use as a single entity that handles both ingestion and transformation of data. Furthermore, the advantages of using a single system to ingest and process the stream are - lower administrative overhead, easier maintenance and lower cost.

The remainder of this paper is organized as follows:

- Section 2 discusses related works in real-time data processing and geofencing.
- Section 3 describes the key features and advantages of using Apache Pulsar.
- Section 4 details the architecture of the system and its various components.
- Section 5 discusses a Case Study of a particular real-time geofencing system and the design considerations made while choosing each of its components.
- Section 6 concludes and provides future directions.

2   Related Work

Wang et al. [8] have proposed a scalable geofencing based system using agricultural machine data to send real-time alerts to farmers. They have proposed a scalable processing system using Apache Kafka to ingest the live stream of data and a cloud Apache Storm architecture that ingests this data and performs the necessary transformation. Since this work has used two separate components for ingestion and processing of data, it takes more time and resources for initial set-up and subsequent maintenance of the pipeline. Google has built a Geolocation Telemetry system [11] to add location-based context to telemetry data using Google Pub/Sub. This is done to ingest the data and python scripts that reverse-geocodes the data to convert latitude and longitude to a street address, calculates the elevation above sea level, and computes the local time offset from UTC by using the timezone for each location. This design is easy and effective for simple stream operations but will not be scalable when the size, velocity and heterogeneity of the stream increases.

Bogdan et al. [7] proposed a geofencing service based on the Complex Event Processing (CEP) paradigm and discussed its applications in the Ambient Assisted Living (AAL) space. The CEP paradigm is very useful when performing intensive stateful transformations by involving pattern detection and generation of secondary pattern streams from the existing ones. However, for simple geofence transformations, such complex pattern detection is not required and hence a dedicated stream processing system which supports CEP is entirely unwarranted. WangJun et al. [9] have proposed a distributed data stream processing pipeline using Apache Flume, Apache Kafka and Apache Spark. In this work, Apache Flume [17] is used as a point-point queue which collects data from multiple sources, Apache Kafka is used for segregating data under different topics and real-time streaming while Apache Spark is used for data transformation and analysis. Nevertheless, when the transformations performed are simple, all these three components can be replaced with a single entity such as Apache Pulsar which supports Data Queuing, Streaming and Simple Processing thus tremendously reducing the overall complexity and maintenance cost of the data pipeline.

3   Apache Pulsar

Apache Pulsar is an open-source distributed pub-sub messaging system which has both stream processing and lightweight computational capabilities. Pulsar comprises a set of brokers, bookies and an inbuilt Apache ZooKeeper [3] for configuration and management. The bookies are from Apache Bookkeeper [4] which provide storage for the messages until they are consumed.

Unlike current messaging systems that have taken the approach of co-locating data processing and data storage on the same cluster nodes or instances, Pulsar takes a cloud-friendly approach by separating the serving and storage layers. Pulsar has a layered architecture with data served by stateless "broker" nodes, while data storage is handled by "bookie" nodes. The architecture of Pulsar is shown in Fig. 1.

3.1   Key Advantages

Low Latency with Durability. Pulsar is designed to have low publish latency (<5 ms) at scale. To confirm this claim, benchmark tests [12] have been performed in literature using the performance analysis tools provided by OpenMessaging Benchmark [14]. These tests were performed with different load sizes and

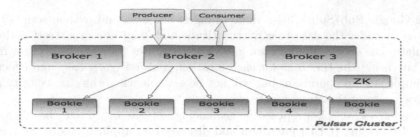

Fig. 1. Pulsar architecture [13]

Table 1. Pulsar Latency Benchmark with 1 topic and 16 partitions [14]

Latency type	Avg	50th	75th	99th	99.99th
Publishing (milliseconds)	2.677	2.655	3.23	3.825	20.883
End-to-end (milliseconds)	3.165	3.0	3.0	12.0	190.0

partitions to analyze the latency of Pulsar. The results of the test are shown in Table 1.

The same test was performed for Apache Kafka to compare its latency against that of Pulsar. The results of this test are tabulated in Table 2.

Table 2. Kafka Latency Benchmark with 1 topic and 16 partitions [14]

Latency type	Avg	50th	75th	99th	99.99th
Publishing (milliseconds)	8.479	8.152	9.64	169.57	211.64
End-to-end (milliseconds)	11.03	10.0	12.0	28.0	259.0

Hence, for latency-sensitive workloads, Pulsar performs better than Kafka. It can provide low latency, as well as strong durability guarantees. This is essential to a system that involves real-time processing and querying, and hence Pulsar outperforms other message queues in these aspects.

Persistent Messaging. Pulsar provides guaranteed message delivery for applications. If a message successfully reaches a Pulsar broker, it will be delivered to its intended target. The non-acknowledged messages are stored durably until they are delivered to and acknowledged by consumers. Pulsar also provides automatic retries until the consumer consumes the messages.

Pulsar Functions. Pulsar Functions are light-weight compute logic that can be easily deployed along with Pulsar to perform real-time data transformation on the stream, thereby eliminating the need for a dedicated stream processing engine. This proves to be very critical for the use case discussed in this paper because it eliminates the necessity to have a separate stream processing logic/system to transform the location coordinates to specific geofenced areas. A Pulsar function can be deployed along with the Pulsar cluster to perform this function efficiently and reliably.

Horizontal Scalability. Pulsar has a layered architecture that enables scaling brokers and bookies independently while maintaining the Zookeeper cluster to coordinate the system. This makes it very easy to horizontally scale the system when the workload increases. This is an essential feature that is required when more devices are tracked, and the message rate increases.

Fault Tolerance. Unlike traditional pub/sub systems, pulsar uses a distributed ledger. Pulsar breaks down the huge logs into several smaller segments, and it distributes those segments across multiple servers while writing the data using Apache Bookkeeper as its storage layer. This makes it easier to add a new server in case of failures, as there is no need to copy the logs of the failed server. Further, since logging is performed in bookkeeper, it reduces the brokers' stress to log data while handling high-velocity real-time data streams, resulting in improved overall performance.

Shared Subscriptions and Partitioned Topics. Pulsar subscriptions have the provision to add as many consumers as needed on a topic, with Pulsar keeping track of all of them. If the consuming application cannot keep up, a shared subscription can be used to distribute the load among multiple consumers. Also, partitioned topics that multiple brokers handle can be used to increase the throughput, thus making it highly scalable.

4   System Architecture

This section illustrates the proposed system for stream handling and processing. The data pipeline for this system is based on streaming architecture. The data pipeline comprises the following modules: Stream handling, Stream Processing, Storage, and Querying. The overall system diagram is shown in Fig. 2.

4.1   Stream Handling

Moving devices transmit their current location coordinates, taken as a streaming input to the pipeline using a publish/subscribe system - Apache Pulsar. The location coordinates are continuously published onto a pulsar topic - a named channel used to pass messages published by producers to consumers who process those messages. Once the data is in a topic, the data is persistently stored (by Apache Bookkeeper) until it is processed.

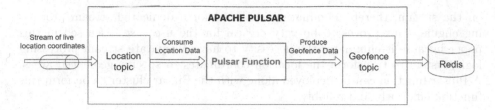

Fig. 2. System diagram

4.2 Stream Processing

The location coordinates present in the topic have to be processed reliably, wherein all the records should be guaranteed to be processed exactly once. The messages are to be processed in the order they are published to maintain the correct context. Since the stream of data is published into the topic continuously, the processing also has to be performed in real-time. Pulsar Functions are capable of performing such transformations with the guarantees mentioned above. Pulsar functions can receive messages from one or more input topics, and every time a message is published to the topic, the function code is executed. For this, a Pulsar function has been written, which reads a message from the location topic, maps the coordinate to its corresponding geofenced area from the list of geofenced areas that the user has provided. Subsequently, the Pulsar Function publishes the transformed geofenced information into a geofence topic. Conversion of the location data to the geofenced area can be realized using Point in Polygon algorithms with the help of R-Trees [2] or Quad Trees to navigate quickly through the search space. Isolating the processing logic to Pulsar Functions makes it easier to adopt the proposed system by easily porting the existing geofencing logic into a corresponding Pulsar Function. Any change to the geofencing logic can be realized by changing the Pulsar Function without affecting any other part of the pipeline. This enhances the portability and pluggability of the overall system.

4.3 Storage and Querying

Data in the geofence topic is updated on a record by record basis in a swift lookup table implemented with Redis [18] or Memcached [19], to find metrics such as the number of devices in a geofenced area, the transition from one geofenced area to another for an individual device and so on, in real-time.

Apart from this, data persistence can be achieved by capturing periodic snapshots of the lookup table and storing it in a persistent database/data warehouse. This can also be realized alternately by using Pulsar IO Connectors to export the data into other databases or messaging systems such as Apache Cassandra [6], Aerospike [20], Apache Kafka, and RabbitMQ, where further transformations can be performed before appropriately storing the data. This data accumulated over a long period can be used for analytics and business intelligence operations to improve decision-making.

In order to present a proof of concept for the proposed pipeline, the following case study has been implemented as a prototype.

5 Case Study - Finding Cab Hotspots Using Real-Time Geofencing

Cab-hailing applications such as Ola or Uber typically use proximity information to assign the nearest cab to every incoming customer request. The lack of restrictions on unoccupied cabs' movement may result in an uneven distribution of cabs across geographical regions and failure to satisfy customer requests. To make optimal use of available cabs, live cab location and customer request data need to be analyzed and cabs need to be redistributed across the various regions based on the identification of hotspots.

The proposed data pipeline has been implemented for this case study application primarily using Apache Pulsar. Other appropriate technologies, such as Redis datastore and Google BigQuery [21] data-warehousing solution have also been deployed to efficiently query and maintain data.

5.1 Design Overview

Cabs belonging to a particular fleet continuously transmit their current location using a hardware device present in them. This stream of cab locations is published into a location topic in Apache Pulsar. Using Pulsar Functions, the published location data is read from the location topic, transformed into geofence data using the point-in-polygon algorithm combined with R-Trees, and subsequently published back into a different Pulsar topic, namely, geofence topic. The geofence data from that topic is consumed by a subscriber and stored in a read-write optimized lookup table. Periodic snapshots are taken from the lookup table and sent to a data warehouse for performing further analytics. The system generates alerts if the number of cabs in a particular geofenced area exceeds a pre-specified threshold. Also, the number of cabs in a specific geofenced area can be queried from the Redis table in real-time. The system diagram is shown in Fig. 3.

5.2 Data Ingestion and Transformation

The cab location data is published continuously in the Pulsar location topic. The location data published is later consumed by Pulsar Functions. The set of geofenced areas under consideration has been used to build an R-Tree. A combination of point-in-polygon algorithm and R-Trees has been used to transform a given location coordinate to its corresponding geofenced area number. This combination helps in quickly narrowing the search space and finding the corresponding geofenced area. This transformation logic has been written into a

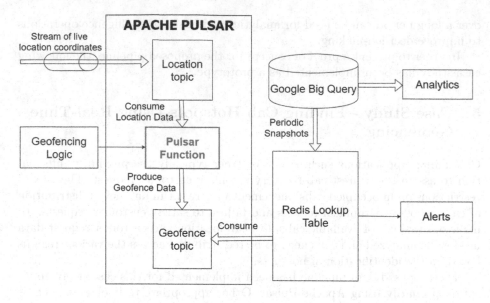

Fig. 3. Cab hotspot detection system

Pulsar Function that takes the location topic as the input and publishes the transformed geofence data into a geofence topic in Pulsar. Thus, the Pulsar function consumes every message published in the location topic, applies the transformation function, and publishes the geofence data in the geofence topic.

5.3   Hotspot Detection

The geofence data published by the Pulsar Functions are consumed by the hotspot detection module, which updates a Redis Table. Whenever geofence data is read from the Pulsar topic, a Redis table corresponding to the number of cabs in the current geofence is updated. This table contains the number of cabs in each geofenced area. Whenever a geofenced area with the number of cabs exceeding the threshold is encountered, an alert message is sent indicating a cab hotspot in that area.

In addition to Redis, Memcached, and MongoDB [22] have also been also evaluated as potential alternatives. However, Redis is more suitable as it is better than Memcached in terms of memory management and handling of free space and has faster disc access time than MongoDB, which is a crucial consideration while designing a real-time system.

5.4   Persistent Storage

Since Redis has volatile memory, snapshots of the Redis table(s) are taken periodically and stored in a data warehouse to have persistent storage. Google Big-Query has been chosen as the data warehouse as it can handle large quantities of data and is easy to set up. As this case study has been carried out for an educational purpose, free availability without any associated cost was an important consideration.

Amazon Redshift [23], LucidDB [24] and PostgreSQL [25] have also been considered as alternatives. Google BigQuery is preferred over these choices as Redshift does not have a free pricing tier, LucidDB was discontinued in 2014, and PostgreSQL is only capable of handling data of the order of a few terabytes. However, Redshift is one of the best cloud data warehousing options when there are no financial constraints.

5.5 Validation of the Data Pipeline

To validate the working of the pipeline, an Azure pulsar cluster across three nodes running Ubuntu 18.04 with 20 GB SSD storage each was set up. The nodes have Intel Xeon CPU E5-2673 and primary memory of 16 GiB each. Java runtime environment was installed on each node.

A stream of data was generated and ingested into the pub/sub system at various rates. Since no real-time streaming source was available, the stream data had to be generated from a static dataset. Chicago cabs dataset [26] was chosen for this purpose. A python message producer program with the pulsar client and producer object was created. The cabs were initialized to random locations in the state of Chicago. Their current locations were stored in the Redis database. Random destination points for each cab were picked. The cabs were made to move from their current location point to a point P calculated as $\frac{currentlatitude+destinationlatitude}{2}, \frac{currentlongitude+destinationlongitude}{2}$. The point P and the destination point were fed into a "Haversine Function" which determines the great-circle distance between two points on a sphere given their longitudes and latitudes. If the haversine distance is above a certain threshold, the process is repeated. Otherwise, it indicates that point the P is very close to the destination, and it is approximated to be the destination. This point becomes the new current location, and a new destination point is chosen as described earlier, and the process is repeated continuously. This generates a continuous stream of cab movement data, which was ingested into Pulsar in real-time. The movement of the cabs has been visualized using a Google Maps API where markers are plotted dynamically. The position of cabs at an instant is shown in Fig. 4.

5.6 Results and Discussions

The performance of the proposed data pipeline has been investigated using a single producer with 2 threads and a message size of 64 bytes to measure the publish latency for different message rates. The results are shown in Table 3.

It can be observed that the change in latency is very negligible, even with a significant increase in the message rate. This proves the robustness and the scalability of the proposed system.

The percentile latency is plotted in Fig. 5 and it is used to monitor the spike in latency under heavy load.

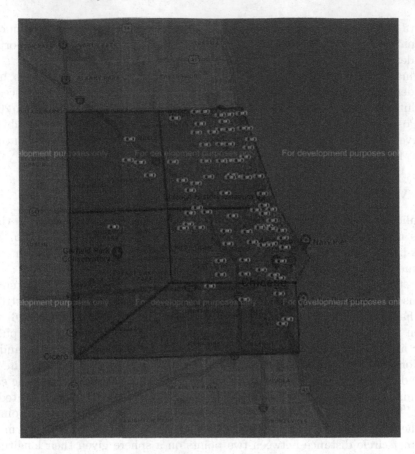

Fig. 4. Cab visualization

Table 3. Message rate vs avg. latency (message size: 64 bytes)

Message rate (msg/sec)	Avg. latency (ms)
10	12.126
100	10.796
1000	26.218
10000	30.285
100000	25.551

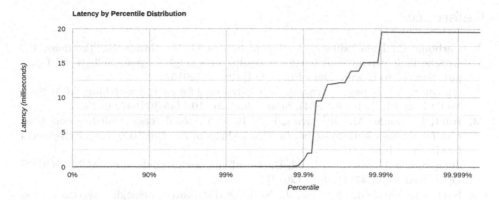

Fig. 5. Latency by percentile distribution (100000 msg/s)

From Fig. 5, it can be seen that the publish latency is less than 1 s for 99.9% of the messages demonstrating the system's ability to handle heavy load while maintaining its performance guarantees. The spike observed in the remaining 0.1% can be reduced either by increasing the broker systems' processing power or by adding additional brokers to the cluster.

Experiments have also been conducted to verify the fault tolerance and message delivery guarantees of the system. Whenever a broker node failed, all the unacknowledged messages were replayed from the bookkeeper once the broker node was back up. All the published messages were consumed without any loss of messages. When running in the once guaranteed processing mode, all the messages published were processed exactly once by the pulsar functions. This shows the capability of the system to recover from failures while still maintaining the context and the processing guarantees.

6 Conclusions

A scalable data pipeline for real-time geofencing using Apache Pulsar has been proposed in this paper. The proposed data pipeline has been designed to handle a high velocity of requests and provide real-time responses. The proposed design is simple and involves fewer components making it easy to set up and maintain. The isolation of the geofencing logic to Pulsar Functions makes it easier to adopt this system with any existing geofencing logic. Usage of Pulsar as the sole stream handling and processing engine helps handle streaming requests with very high velocity and reliably process all the messages. Pulsar is highly fault-tolerant because of its 3-Tier architecture and distributed logging using Apache Bookkeeper, making the system scalable and robust without compromising real-time responsiveness.

The case study performed corroborates the claims made. It contains useful insights into various design considerations when choosing other systems to work in combination with the proposed data pipeline. Thus, it could serve as a reference for future works involving the proposed data pipeline.

References

1. Carbone, P., Katsifodimos, A., Ewen, S., Markl, V., Haridi, S., Tzoumas, K.: Apache flink: stream and batch processing in a single engine. Bull. IEEE Comput. Soc. Tech. Comm. Data Eng. **36**(4), 28–38 (2015)
2. Guttman, A.: R-trees: a dynamic index structure for spatial searching. ACM SIGMOD Rec. **14**(2), 1–11 (1984). https://doi.org/10.1145/971697.602266
3. Hunt, P., Konar, M., Junqueira, F.P., Reed, B.: ZooKeeper: wait-free coordination for internet-scale systems. In: Proceedings of the USENIX Annual Technical Conference, June 2010
4. Junqueira, F.P., Kelly, I., Reed, B.: Durability with BookKeeper. ACM SIGOPS Oper. Syst. Rev. **47**(1), 9–15 (2013)
5. Kreps, J., Narkhede, N., Rao, J.: Kafka: a distributed messaging system for log processing. In: Proceedings of the NetDB, pp. 1–7 (2011)
6. Lakshman, A., Malik, P.: Cassandra: a decentralized structured storage system. ACM SIGOPS Oper. Syst. Rev. **44**(2), 35–40 (2010)
7. Târnaucă, B., Puiu, D., Nechifor, S., Comnac, V.: Using complex event processing for implementing a geofencing service. In: IEEE 11th International Symposium on Intelligent Systems and Informatics, Subotica, Serbia, pp. 1–6, September 2013
8. Wang, Y., et al.: An open-source infrastructure for real-time automatic agricultural machine data processing. In: 2017 ASABE Annual International Meeting, Spokane, Washington, pp. 1–13, July 2017. https://doi.org/10.13031/aim.201701022
9. Wang, J., Wang, W., Chen, R.: Distributed data streams processing based on Flume/Kafka/Spark. In: 2015 3rd International Conference on Mechatronics and Industrial Informatics, Zhuhai, China, pp. 948–952, October 2015. https://doi.org/10.2991/icmii-15.2015.167
10. Zaharia, M., Chowdhury, M., Franklin, M.J., Shenker, S., Stoica, I.: Spark: cluster computing with working sets. In: Proceedings of the 2nd USENIX Conference on Hot Topics in Cloud Computing, HotCloud 2010, Berkeley, CA, USA, p. 10 (2010)
11. Building a Scalable Geolocation Telemetry System Using the Maps API. https://cloud.google.com/solutions/scalable-geolocation-telemetry-system-using-maps-api. Accessed 9 Aug 2020
12. Kafkaesque. https://kafkaesque.io/performance-comparison-between-apache-pulsar-and-kafka-latency/. Accessed 4 Jan 2021
13. Pulsar Architecture. https://pulsar.apache.org/docs/en/concepts-architecture-overview/. Accessed 4 Jan 2021
14. OpenMessaging Benchmark. http://openmessaging.cloud/. Accessed 4 Jan 2021
15. RabbitMQ. https://www.rabbitmq.com/documentation.html. Accessed 4 Jan 2021
16. Apache Pulsar. https://pulsar.apache.org/. Accessed 4 Jan 2021
17. Apache Flume. https://flume.apache.org/. Accessed 4 Jan 2021
18. Redis. https://redis.io/. Accessed 4 Jan 2021
19. Memcached. https://memcached.org/. Accessed 4 Jan 2021
20. Aerospike. https://www.aerospike.com/. Accessed 4 Jan 2021
21. Google BigQuery. https://cloud.google.com/bigquery. Accessed 4 Jan 2021
22. MongoDB. https://www.mongodb.com/. Accessed 4 Jan 2021
23. Amazon Redshift. https://aws.amazon.com/redshift/. Accessed 4 Jan 2021
24. LucidDB. https://dbdb.io/db/luciddb. Accessed 4 Jan 2021
25. PostgreSQL. https://www.postgresql.org/. Accessed 4 Jan 2021
26. Chicago Taxi Trips. https://www.kaggle.com/chicago/chicago-taxi-trips-bq/. Accessed 4 Jan 2021

Crop Recommendation by Analysing the Soil Nutrients Using Machine Learning Techniques: A Study

Vaishnavi Jayaraman[1], Saravanan Parthasarathy[1], Arun Raj Lakshminarayanan[1(✉)], and S. Sridevi[2]

[1] B. S. Abdur Rahman Crescent Institute of Science and Technology, Chennai, India
{saravanan_cse_2019,arunraj}@crescent.education
[2] Vels Institute of Science, Technology and Advanced Studies, Chennai, India
sridevis.se@velsuniv.ac.in

Abstract. According to India Brand Equity Foundation (IBEF), 32% of the global food market is dependent on Indian agricultural sector. Due to urbanisation, the fertile land have been utilised for non-agricultural purposes. The loss of agricultural lands impacts the productivity and results with diminishing yield. Soil is the most important factor for the thriving agriculture, since it contains the essential nutrients. The food production could be improved through the viable usage of soil nutrients. To identify the soil nutrients, the physical, chemical and biological parameters were examined using many machine learning algorithms. However, the environmental factors such as sunlight, temperature, humidity, and rainfall plays a major role in improving the soil nutrients since it is responsible for the process of photosynthesis, germination, and saturation. The objective is to determine the soil nutrient level by accessing the associative properties including the environmental variables. The proposed system termed as Agrarian application which recommends crops for the particular land using classification algorithms and predicts the yield rate by employing regression techniques. The application will help the farmers in selecting the crops based on the soil nutrient content, environmental factors and predicts the yield rate for the same.

Keywords: Soil nutrients · Environmental factors · Machine learning · Prediction · Crop recommendation system · Yield rate

1 Introduction

Fertile soil deposited by the rivers is a plinth of ancient civilizations. In the initial years, the availability of nutrient soil and abundant water in delta regions motivate humans to perform agricultural practices. The nutrient enriched alluvial soil boosted the plant growth and increased the yield. Since humans are connected to the land through agriculture, the soil nutrients played a key role in anthropological evolution which leads to cultural progress. In India, about 52% of the people rely on agricultural industry and

© IFIP International Federation for Information Processing 2021
Published by Springer Nature Switzerland AG 2021
V. Krishnamurthy et al. (Eds.): ICCIDS 2021, IFIP AICT 611, pp. 15–26, 2021.
https://doi.org/10.1007/978-3-030-92600-7_2

it contributes around 18% of GDP. According to 'The Energy and Resource Institute' (TERI) [1], about 2.5% of Indian economic system relies on soil quality. The physical, chemical, and biological properties of soil determine the agricultural outcome. The soil textures, porosity, capacity of holding water, pH, Electrical Conductivity (EC), Organic Carbon (OC) are some of the physical and chemical properties of the soil. Since most of the nutrients are soluble in acidic soil, the pH value is an important attribute in nutrient management [2]. EC is used to calibrate the soil salinity which integrated the macro and micronutrients. In general, 2–10% of the total dry weight of the soil is covered by OC. The decayed animals, plants, and microorganisms exhibit OC, which is ingested by the crops. The mineral and non-mineral nutrients were determined by the biological properties [3]. The mineral nutrients can be classified into macronutrients, secondary nutrients, and micronutrients (Table 1).

Table 1. List of mineral nutrients

Nutrient type	Nutrients	Symbol
Macronutrients	Nitrogen	N
	Phosphorous	P
	Potassium	K
Secondary nutrients	Calcium	Ca
	Magnesium	Mg
	Sulphur	S
Micronutrients	Iron	Fe
	Manganese	Mn
	Zinc	Zn
	Copper	Cu
	Boron	B
	Molybdenum	Mo
	Chlorine	Cl
	Nickel	Ni

Nitrogen (N) is the most widely exploited elementary unit which holds all the proteins and the component of chlorophyll. The saplings acquired the Nitrogen either in the form of ammonium or nitrate. Phosphorus (P) is the least utilized supplement that is enriched with the nucleic acid. It strengthens the root, which is absorbed as orthophosphate ions. The soil pH and the ratio of orthophosphate are inversely proportional. Potassium (K) is essential for activating the enzymes and regulating osmosis. The secondary nutrients are required in a smaller quantity when compared to the macronutrients. Calcium (Ca), Magnesium (Mg), and Sulphur (S) is responsible for fortifying the cell wall, maintaining the electrical balance and enriching the amino acids respectively. Even though the requirements of micronutrients are less than 1%, they make a significant impact on plant growth and agricultural turnout. The micronutrients such as Zn, Mg, Cu, and Fe induce the activation of enzymes, whereas the Fe, Zn, and Mg are also allied to chlorophyll. The non-mineral elements such as Carbon (C), Oxygen (O), and Hydrogen (H) are obtained through the air (CO_2) and water (H_2O). Though the utilization level of nutrition varies, every nutrient plays an imperative part in keeping soil fertility. The factors including erosion, leaching, and continuous farming affect the soil nutrients and result in a reduction of productivity. Organic and inorganic fertilizers are used to tackle this situation. On the contrary, the excessive use of fertilizers leads to the loss of natural replenishment of soil. The traditional laboratory-based soil analysis required plenty of time and resources. Instead, the Machine Learning Techniques are competent to accost these issues with minimal time and cost. The applicability and advancements of Machine Learning Techniques were discussed in this study.

2 Literature Survey

Suchitra et al., predicts the soil fertility index and soil pH [4] using Extreme Learning Machine Algorithm (ELM). The fertility index defines the quality of soil to ensure plant growth and the pH is to measure the acidity or alkalinity of the same. The data were collected from various farming lands in the state of Kerala. The physical and chemical parameters of the samples were examined in the soil laboratory. The Extreme Learning Machine Algorithm was implemented with various activation functions such as triangular basis, sine-squared, hard limit, Gaussian Radial Basis (GRB), and hyperbolic tangent. ELM-GRB performs best with accuracy of 80%, in calibrating the soil fertility index by considering both the accuracy and kappa. ELM-hyperbolic achieves accuracy of 75%, for predicting the soil pH. Zonlehoua Coulibali et al. proposed an approach that predicts the optimal macronutrient (NPK) requirements [5]. They were usually determined by the soil type, weather, and many other variables. The dataset for the crop potato was collected from Quebec, Canada. The key features were extracted by using the Extra Tree Regressor function. An optimum model was developed and correlated from KNN, Gaussian model, Random Forest, Mitscherlich model, and Neural Network. Since its R^2 value is between 0.60 and 0.69, the Gaussian model predicted the macronutrient level in a better way than the other models.

The better nutrient management would result in the production of first grade Peach fruit. D. Betemps et al. suggests that the above-mentioned outcome could be achieved through Humboldtian Diagnosis [6] and Machine Learning Techniques such as Random Forest, Neural Networks, Support Vector Machine, and Stochastic Gradient Descent. In this study, Random Forest predicts the soil nutrients effectively with 80% of accuracy. H. Mollenhorst et al. employed a couple of Machine Learning algorithms to predict the soil phosphorus with the application of dairy farm manure [7]. A historic dataset for the past 24 years was collected from the farm DeMark, Netherlands. Gradient Boosting Machine (GBM) and Decision Tree were engaged for predicting the phosphorus content before the application of the first manure. The GBM model outperforms the other one, with an RMSE value between 7.33 and 8.22.

A Decision Support System (DSS) for the proper usage of macronutrient fertilizer was developed by R. Meza-Palacios et al., to enhance sustainability [8]. The physical and chemical properties of the soil were analysed in the laboratory. The key variables chosen were Electrical Conductivity, soil organic matter, and soil texture. The DSS consists of two fuzzy models; the Edaphic condition model (EDC) and the NPK fertilization model.

The physical and chemical index of the soil was controlled by the EDC model. The outcome of EDC, available macronutrients, and yield rate was controlled by the NPK model. The R^2 rates of macronutrients (NPK) shown in DSS were 0.981, 0.9702, and 0.9691. Thus, the DSS model suggests NPK accurately.

Chunyan Wu et al., recommended the required soil quality and nutrient contents of Dacrydium pectinatum [9] in China. In this study, six Support Vector Machine (SVM) models, and four Neural Networks (NN) models were utilised. The soil was collected from the four corners and center part of the field. The sample had been examined in a laboratory using the Kjeldahl method. The macronutrients and organic matter present in the soil were considered in the analysis. The adapted SVM models were Local Mixture-based Support Vector Machine (LMSVM), Fast local kernel Support Vector Machine (FSVM), Proximal Support Vector Machine (PSVM), Localized Support Vector Machine (LSVM), KNN and SVM integrated algorithms (SVM-KNN and KNN-SVM). The NN models, applied were Back-Propagation Neural Network (BPNN), Field Probing Neural Network (FPNN), MultiLayer-Propagation feed-forward Neural Network (MLPNN), Generalized Regression Neural Network (GRNN). From all these 10 models GRNN model with the least RMSE value performs best in the NN model and KNN-SVM (partial SVM) out-performs all the other SVM models with accuracy of 95%.

The Walkley-Back method [10] of soil examination was done by K. John et al., to estimate the soil organic compound in alluvial soil with specific parameters. The predictors considered were clay index, base saturation, Normalized Difference Moisture Index (NDMI), Land Surface Temperature (LST), Normalized Difference Built-Up Index (NDBI), Soil Adjusted Vegetation Index (SAVI), Normalized Difference Vegetation Index (NDVI), and Ratio Vegetation Index (RVI). The digital mapping of soil with the environmental variables was also implemented with the Cubist Regression, ANN, Multi-Linear Regression (MLR), Random Forest, and SVM. The Random Forest performed better than other algorithms with the R^2 value of 0.68.

Chunyan Wu et al., identified the features of topsoil nutrients and biomass in the dark brown soil of Northeast forest in China. To predict the above-ground biomass [11], Chunyan Wu utilized various ML models, out of which GRNN brought a higher accuracy, with a gradient of 0.937. The adapted NN algorithms were the Group Method Of Data Handling (GMDH), Artificial Neural Network (ANN), Adaptive Neuro-Fuzzy Inference System (ANFIS), Generalized Regression Neural Network (GRNN), and Support Vector Machine (SVM). The result shows that if the depth of the soil increases, soil nutrients decrease. M. Shahhosseini et. al, aimed to predict the nitrate loss using Agriculture Production Systems SiMulator Cropping System (APSIM) model [12] with historical data from US Midwest. To assist the subsequent progress of the decision support tool the four ML algorithms such as Extreme Gradient Boosting, LASSO Regression, Random Forest, Ridge Regression and their ensembles were executed as metamodels for APSIM. Random Forest predicts the nitrate loss more precisely, with an RRMSE of 54%.

C. Ransom et. al, proposed an approach to optimize the economically optimal N rates (EONR) using eight statistical and ML approaches [13]. The statistical and ML approaches applied were Stepwise Regression, Ridge Regression, Least Absolute Shrinkage, And Selection Operator (LASSO), Elastic Net Regression, Principal Component Regression (PCR), Partial Least Squares Regression (PLSR), Decision Tree and

Random Forest have been implemented to design an Nitrogen recommendation tool. The ML model that balanced the Nitrogen recommendation tool was Random Forest with an inclined R^2 value between 0.72 and 0.84. The RMSE value lied between 41 and 94 kgNha−1. However, the Decision Tree dealt with a minimal quantity of parameters, and results with an inclined R^2 value between 0.15 and 0.51. The RMSE value of this model lied between 16 and 66 kgNha−1.

J. Massah et al., used the cone penetrometer [14] to examine the sample soil which contains rotten matter, decayed leaves, and water, to estimate the modification in soil penetration resistance (SPR). The SPR was analysed using the following algorithms KNN, SVM, ANN, Random Forest, Levenberg-Marquardt backpropagation ANN, Naïve Bayes, and Decision Tree. SVM-Gaussian kernel achieves higher accuracy in forecasting the modifications in SPR, with the R^2 value of 0.982 and the mean square error (MSE) value between 0.02 and 0.09. Soft sensors based on DBN-ELM (Data Belief Network-Extreme Learning Machine) measuring the nutrients in the water content of soil-free cultivation [15]. The parameters considered were EC, pH, Circulation speed, and temperature. RMSE values of Least Square LS, ELM, and DBN-ELM are 2.3877, 1.7838, and 1.2414 respectively. From the derived RMSE value, the integrated ML model DBN-ELM performed better than the others.

M.S. Sirsat proposes et al., a village-wise prediction of soil fertility index for micronutrients [16]. The soil samples of ten villages were collected from the state of Maharashtra. The Neural networks, Extreme Randomised Regression, Profound Learning, LASSO, Ridge Regression (RR), Bayesian model, Support Vector Regression, and Random Forest were the models used in this examination. Extreme Randomised Regression Trees out-performed the others through their fastness, with the accuracy rate of 97.5%. Yuefen Li et al., proposed a machine learning approach to identify the other nutrients mandatory for plant growth other than Carbon, Phosphorus, and Potassium [17]. The proposed approach also aimed to predict the nutrients present in the soil. A couple of algorithms such as Radial Basis Function Neural Network (RBFNN) and SVM, were used to predict the nutrient content. The overall performance of both SVM and RBFNN models predicts the soil nutrients in an efficient manner with a prediction accuracy of 99.85% and 98.45% respectively.

Multiple statistical techniques were employed by M. Hosseini et al., to predict the soil Phosphorus (P) using the ML models. The employed statistical techniques were fuzzy inference system, Adaptive neuro-fuzzy inference system, Regressions, ANN, Partial Least Square Regression (PLSR), and Genetic Algorithms (GA) [18]. ANN forecasts the soil Phosphorus level better than other algorithms with the R^2 rate of 0.912 and RMSE rate of 4.019, by considering the soil organic matter and pH. To predict the soil phosphorus, a couple of models GA and PLS came up with statistical formulas to find the best fit. A comprehensive model based on SVM was designed to classify the soil quality [19]. The soil samples were collected from Taiyuan city, the heavy metals such as Cadmium, lead, chromium, and Nickel were chemically determined, to find the combination of contaminated soil and quality soil. Using this correlated dataset, the designed SVM model predicts the soil quality with accuracy of 98.33%.

A decision model based on a fuzzy Bayesian approach was devised to predict the soil fertility level, which would help in selecting the paddy variety [20]. The sample soil was collected from various farming lands in Vellore and was tested in the district soil test laboratory. The attributes considered were EC, pH, N, P, K, Zn, Fe, Cu, and Mn. The study identified the soil quality and suggested the paddy type. The fuzzy Bayesian approach achieved the accuracy of 90.2%. Hao Li et. al, evaluated the nutrient content of the soil and using a pair of ML models. GRNN, SVM, and MLR [21] were used. The prediction accuracy of SVM and MLR was 77.87% and 83.00% respectively, whereas the accuracy achieved by GRNN was 92.86%. Hence, GRNN predicts the soil nutrients in a better way a minimal error rate (MSE = 0.27) (Table 2).

Table 2. Key findings and interpretations

S. No	Author(s)	Methodology used	Outcome	Advantage	Disadvantage
1	Suchithra, M. S. and Pai, M. L. (2020) [4]	ELM with various activation functions: tri- angular basis, sine-squared, hard limit, Gaussian radial basis, and hyperbolic tangent	Soil fertility index- ELM-GRB - accuracy 80% Soil pH - ELM-hyperbolic -accuracy 75%	Very fast, efficient and saves time in analysing the soil	The other soil nutrients such as N_2O, P_2O_5, and K_2O were not considered
2	Coulibali, Z., Cambouris, A. N. and Parent, S. É. (2020) [5]	KNN, Gaussian model, Random Forest, Mitscherlich model, and NN	Macronutrient level - Gaussian model with R^2 value between 0.60 and 0.69	Effective performance	The size of dataset is minuscule
3	Betemps, D. L., De Paula, B. V., Parent, S. É., Galarça, S. P., Mayer, N. A., Marodin, G. A. B., Rozane, D. E., Natale, W., Melo, G. W. B., Parent, L. E. and Brunetto, G. (2020) [6]	Random Forest, Neural Networks, SVM, and Stochastic Gradient Descent	Soil nutrient level - Random Forest predicts effectively with 80% of accuracy	Good performance with a minimal number of errors	The assessment was against minimal data points
4	Mollenhorst, H., de Haan, M. H. A., Oenema, J. and Kamphuis, C. (2020) [7]	Gradient Boosting Machine and Decision Tree	Soil Phosphorus content - GBM model outperforms with an RMSE value between 7.33 and 8.22	A closer coherence between the application of P and yield will lead to a great profit	This method is applicable only for sandy soil and almost impossible to scale up

(continued)

Table 2. (*continued*)

S. No	Author(s)	Methodology used	Outcome	Advantage	Disadvantage
5	Meza-Palacios, R., Aguilar-Lasserre, A. A., Morales-Mendoza, L. F., Rico-Contreras, J. O., Sánchez-Medel, L. H. and Fernández-Lambert, G. (2020) [8]	DSS and a pair of fuzzy models; Edaphic condition model (EDC) and NPK fertilization model	NPK level - DSS model suggests NPK with R^2 value 0.981, 0.9702, and 0.9691	The designed tool will perform better for the following crops maize, avocado, grasses, sorghum	The efficiency of the system relied on the diversification of data. The unavailability led to inefficiency
6	Wu, C., Chen, Y., Hong, X., Liu, Z. and Peng, C. (2020) [9]	LMSVM, FSVM, PSVM, LSVM, SVM-KNN, KNN-SVM, BPNN, FPNN, MLPNN, and GRNN	Soil quality -GRNN and KNN-SVM outperforms the other algorithms	Higher prediction accuracy and convergence rate	Uncertainty in generating new models and data acquisitions
7	John, K., Isong, I. A., Kebonye, N. M., Ayito, E. O., Agyeman, P. C. and Afu, S. M. (2020) [10]	Cubist regression, ANN, MLR, Random Forest, and SVM	Organic Compound - Random Forest performed better with the R^2 value of 0.68	Better nutrient management	The other physical parameters had not been considered along with the soil types
8	Wu, C., Chen, Y., Hong, X., Liu, Z. and Peng, C. (2020) [11]	GMDH, ANN, ANFIS, GRNN, and SVM	Soil Nutrients -GRNN brought a higher accuracy, with a gradient of 0.937	The relatively best model with higher efficiency	Lack of effective driving attributes
9	Shahhosseini, M., Martinez-Feria, R. A., Hu, G. and Archontoulis, S. V. (2019) [12]	Extreme Gradient Boosting, LASSO Regression, Random Forest, RR and their ensembles	Nitrate loss - Random Forest predicts more precisely, with an RRMSE of 54%	An impressive result was obtained using a small dataset	The greater exception will lead to a high chance of errors
10	Ransom, C. J., Kitchen, N. R., Camberato, J. J., Carter, P. R., Ferguson, R. B., Fernández, F. G., Franzen, D. W. Laboski, C. A. M., (2019) [13]	Stepwise regression, ridge regression, LASSO, elastic net regression, PCR, PLSR, Decision Tree, and Random Forest	Nitrogen recommendation -Random Forest and Decision Tree with an inclined R^2 value	Achieved greater accuracy with a minimum number of variables	Auxiliary parameters such as soil pattern, weather, and plant genetics had not been taken into account

(*continued*)

Table 2. (*continued*)

S. No	Author(s)	Methodology used	Outcome	Advantage	Disadvantage
11	Massah, J., Asefpour Vakilian, K. and Torktaz, S. (2019) [14]	KNN, SVM, ANN, Random Forest, Levenberg-Marquardt back propagation ANN, Naïve Bayes and Decision Tree	SPR – SVM Gaussian kernel achieves higher accuracy with the R^2 value of 0.982 and MSE value between 0.02 and 0.09	SPR prediction using ML models acts as an alternative to instruments like cone penetrometer	The model only dealt with the data collected from one particular region
12	Wang, X., Hu, W., Li, K., Song, L. and Song, L. (2019) [15]	LS, ELM and DBN-ELM	Water Content - DBN-ELM performed better with an RMSE value of 2.3877, 1.7838, and 1.2414	Excellent feature extraction and higher accuracy	Deliberate performance
13	Sirsat, M. S., Cernadas, E., Fernández-Delgado, M. and Barro, S. (2018) [16]	NN, Extreme randomised regression, Deep Learning, LASSO, Ridge Regression, Bayesian model, Support Vector Regression, and Random Forest	Soil fertlity index - Extreme Randomised Regression Trees out-performed with the accuracy rate of 97.5%	Relatively fast and high performance	Macronutrients and secondary nutrients were not contemplated
14	Li, Y., Liang, S., Zhao, Y., Li, W. and Wang, Y. (2017) [17]	RBFNN and SVM	Soil Nutrients -SVM and RBFNN produced the accuracy as 99.85% and 98.45% respectively	Fast performance, higher credibility, and precision	Particular type of soil and plant alone was considered
15	Hosseini, M., Rajabi Agereh, S., Khaledian, Y., Jafarzadeh Zoghalchali, H., Brevik, E. C. and Movahedi Naeini, S. A. R. (2017) [18]	Fuzzy inference system, Adaptive neuro-fuzzy inference system, Regressions, ANN, PLSR and GA	Soil Phosphorus -ANN forecasts better with the R^2 rate of 0.912 and RMSE rate of 4.019	Simple, fast, and effective performance	The possibility of Empirical error occurrence is high

(*continued*)

Table 2. (*continued*)

S. No	Author(s)	Methodology used	Outcome	Advantage	Disadvantage
16	Liu, Y., Wang, H., Zhang, H. and Liber, K. (2016) [19]	SVM based soil quality model	Soil quality level-the designed SVM model achieved accuracy of 98.33%	A feasible and reliable model	Sensitive to outliers and overfits
17	Lavanya, K., Saleem Durai, M. A. and Iyengar, N. C. S. N. (2015) [20]	Fuzzy Bayesian, Naïve Bayesian, and Neural networks	Soil fertility level - fuzzy Bayesian approach achieved the accuracy of 90.2%	Suitable for both rich and poor datasets	The environmental variables and secondary nutrients were not included
18	Li, H., Leng, W., Zhou, Y., Chen, F., Xiu, Z. and Yang, D. (2014) [21]	GRNN, SVM, and MLR	Soil nutrients -GRNN achieved accuracy of 92.86% with a MSE of 0.27	Effective and fast performance	Fluctuation is relatively lower and pattern recognition had not been considered

3 Research Gap

Fertile soil is not only the combination of essential nutrients alone, the other environmental factors such as sunlight, temperature, humidity, and rainfall, should also be considered. However, the majority of research works concentrated on predicting the level of nutrients to optimize the production. The physical, chemical and biological properties would differ to a great extent with respect to the environmental factors [22]. Soil nutrient management could be achieved through the sustainable agricultural practices. The main objective of the proposed system is to identify the soil nutrient level and assess the same with environmental factors. The consolidated dataset will contain physical, chemical, biological, and environmental factors, whereas the existing system, examines any one of the above factors for predicting the crop yield. The proposed Agrarian application equipped with the state-of-art classification and regression algorithms. The system would also employ the correlation techniques to find the relationship between the crop growth and the other factors. It would help in suggesting the crop for a particular farm and predict the yield rate (Fig. 1).

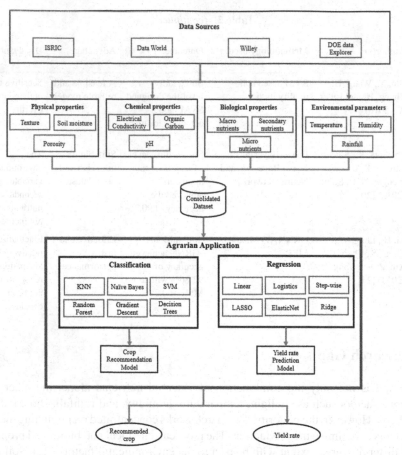

Fig. 1. The proposed Agrarian application workflow

4   Conclusion and Future Work

The nutrient content of the soil is a key factor for the agricultural outcome. The presence of sixteen essential nutrients is necessary for the optimal outcome along with the environmental parameters. Most of the research studies concentrates on evaluating the total amount of nutrients, physical and biological characteristics of soil to optimize the crop yield rate. However, the environmental factors have a higher influence on the level of soil nutrients. The sustainability of the soil could only be ensured by maintaining the proper environmental conditions. The objective of this study is to design a model that suggests the suitable crop for the particular farm using the state-of-art classification algorithms and predicts the yield rate of the recommended crop by employing regression techniques. The proposed Agrarian model would help the farmers in selecting the crops which is expected to give optimum yield, without taking soil tests in laboratories. This work could be further extended by implementing ANN, into the system. Since the ANN is embodied with actvation functions, it is capable of learning any dynamic input-output interaction. Hence, ANN would be employed to improve the soil fertility prediction and recommend the fertilizers based on the variation in environmental factors.

References

1. https://www.teriin.org/project/india-prepares-host-un-conference-curb-land-degradation
2. https://www.esf.edu/pubprog/brochure/soilph/soilph.htm
3. http://www.fao.org/3/a-a0443e.pdfs
4. Suchithra, M.S., Pai, M.L.: Improving the prediction accuracy of soil nutrient classification by optimizing extreme learning machine parameters. Inf. Process. Agric. **7**(1), 72–82 (2020)
5. Coulibali, Z., Cambouris, A.N., Parent, S.É.: Site-specific machine learning predictive fertilization models for potato crops in Eastern Canada. PLoS ONE **15**(8), e0230888 (2020)
6. Betemps, D.L., et al.: Humboldtian diagnosis of peach tree (prunus persica) nutrition using machine-learning and compositional methods. Agronomy **10**(6), 900 (2020)
7. Mollenhorst, H., de Haan, M.H.A., Oenema, J., Kamphuis, C.: Field and crop specific manure application on a dairy farm based on historical data and machine learning. Comput. Electron. Agric. **175**, 105599 (2020)
8. Meza-Palacios, R., Aguilar-Lasserre, A.A., Morales-Mendoza, L.F., Rico-Contreras, J.O., Sánchez-Medel, L.H., Fernández-Lambert, G.: Decision support system for NPK fertilization: a solution method for minimizing the impact on human health, climate change, ecosystem quality and resources. J. Environ. Sci. Health Part A Toxic/Hazard. Subst. Environ. Eng. **55**(11), 1267–1282 (2020)
9. Wu, C., Chen, Y., Hong, X., Liu, Z., Peng, C.: Evaluating soil nutrients of Dacrydium pectinatum in China using machine learning techniques. For. Ecosyst. **7**(1) (2020). https://doi.org/10.1186/s40663-020-00232-5
10. John, K., Isong, I.A., Kebonye, N.M., Ayito, E.O., Agyeman, P.C., Afu, S.M.: Using machine learning algorithms to estimate soil organic carbon variability with environmental variables and soil nutrient indicators in an alluvial soil. Land **9**(12), 1–20 (2020)
11. Wu, C., Pang, L., Jiang, J., An, M., Yang, Y.: Machine learning model for revealing the characteristics of soil nutrients and aboveground biomass of Northeast Forest, China. Nat. Environ. Pollut. Technol. **19**(2), 481–492 (2020)
12. Shahhosseini, M., Martinez-Feria, R.A., Hu, G., Archontoulis, S.V.: Maize yield and nitrate loss prediction with machine learning algorithms. ArXiv (2019)
13. Ransom, C.J., et al.: Statistical and machine learning methods evaluated for incorporating soil and weather into corn nitrogen recommendations. Comput. Electron. Agric. **164**, 104872 (2019)
14. Massah, J., Asefpour Vakilian, K., Torktaz, S.: Supervised machine learning algorithms can predict penetration resistance in mineral-fertilized soils. Commun. Soil Sci. Plant Anal. **50**(17), 2169–2177 (2019)
15. Wang, X., Hu, W., Li, K., Song, L., Song, L.: Modeling of soft sensor based on DBN-ELM and its application in measurement of nutrient solution composition for soilless culture. In: Proceedings of 2018 IEEE International Conference of Safety Produce Informatization, IICSPI 2018, pp. 93–97 (2019)
16. Sirsat, M.S., Cernadas, E., Fernández-Delgado, M., Barro, S.: Automatic prediction of village-wise soil fertility for several nutrients in India using a wide range of regression methods. Comput. Electron. Agric. **154**, 120–133 (2018)
17. Li, Y., Liang, S., Zhao, Y., Li, W., Wang, Y.: Machine learning for the prediction of L. chinensis carbon, nitrogen and phosphorus contents and understanding of mechanisms underlying grassland degradation. J. Environ. Manag. **192**, 116–123 (2017)

18. Hosseini, M., Rajabi Agereh, S., Khaledian, Y., Jafarzadeh Zoghalchali, H., Brevik, E.C., Movahedi Naeini, S.A.R.: Comparison of multiple statistical techniques to predict soil phosphorus. Appl. Soil. Ecol. **114**, 123–131 (2017)
19. Liu, Y., Wang, H., Zhang, H., Liber, K.: A comprehensive support vector machine-based classification model for soil quality assessment. Soil and Tillage Res. **155**, 19–26 (2016)
20. Lavanya, K., Saleem Durai, M.A., Iyengar, N.C.S.N.: Site specific soil fertility ranking and seasonal paddy variety selection: an intuitionistic fuzzy rough set and fuzzy Bayesian based decision model. Int. J. Multimed. Ubiquit. Eng. **10**(6), 311–328 (2015)
21. Li, H., Leng, W., Zhou, Y., Chen, F., Xiu, Z., Yang, D.: Evaluation models for soil nutrients based on support vector machines and artificial neural networks. Sci. World J. **2014**, 1–8 (2014)
22. Fageria, N.K., Baligar, V.C.: Nutrient availability. Encycl. Soils Environ. **4**, 63–71 (2004)

Predicting Customer Churn in Banking Based on Data Mining Techniques

Wafaa A. Alsubaie⬤, Haya Z. Albishi⬤, Khloud A. Aljoufi⬤,
Wedyan S. Alghamdi⬤, and Eyman A. Alyahyan(✉)⬤

Computer Science Department, College of Science and Humanities, Imam Abdulrahman Bin
Faisal University, P.O. Box 31961, Jubail, Kingdom of Saudi Arabia
eaalyahyan@iau.edu.sa

Abstract. One of the most critical challenges facing banking institutions is customer churn, as it dramatically affects a bank's profits and reputation. Therefore, banks use customer churn forecasting methods when selecting the necessary measures to reduce the impact of this problem. This study applied data mining techniques to predict customer churn in the banking sector using three different classification algorithms, namely: decision tree (J48), random forest (RF), and neural network (MLP) using WEKA. The Results showed that J48 had an overall superior performance over the five performance measures, compared to other algorithms using the 10-fold cross-validation. Additionally, the InfoGain and correlation features selection method was used to identify significant features to predict customer churn. The experiment revealed that both algorithms work better when all features are utilized. In short, the results obtained help in predicting which customer is likely to leave the bank. Furthermore, identifying these essential features will help banks keep customers from churning out and compete with rival banks.

Keywords: Classification · Customer churn · Churn prediction · Data mining

1 Introduction

The tremendous development in various aspects of life has led to an increase in the growth of data in its various forms, which has increased the need for smart tools and techniques to reduce the efforts required in the search for useful information. In turn, it has helped to extract new and useful knowledge, also known as data mining techniques [1]. The same is the case in the banking industry, as it requires these technologies to improve and support their decisions in all of their activities when historically analyzing the activities of their customers [2]. One of the biggest banking concerns is customer churn, and they are waging a fierce war to preserve and gain customer loyalty [3]. Customer churn is the loss of custom when individuals change from one provider to another. The problem of customer churn can be reduced by extracting knowledge from the data that helps classify customer behavior, identifying the most important features that affect customer churn, using data mining techniques, and uncovering hidden behaviors

© IFIP International Federation for Information Processing 2021
Published by Springer Nature Switzerland AG 2021
V. Krishnamurthy et al. (Eds.): ICCIDS 2021, IFIP AICT 611, pp. 27–39, 2021.
https://doi.org/10.1007/978-3-030-92600-7_3

in data sets that may not have been visible before. Previous studies have shown the significance of predicting customers churn at an early stage to avoid losing customers, as this can have a significant effect on a bank's profits and be costly. To the researchers' knowledge, studies that used the Kaggle data set to address a customer churn problem had not previously used Neural Network (MLP), Random Forest, and Decision Trees (J48) algorithms together and compared them in terms of which performs best. Therefore, in this study, a machine learning model was built to predict customer churn in the banking sector using Neural network (MLP), random forests, and decision trees (J48), taking into consideration the optimization strategies for all models. This study's results will undeniably be beneficial to those who wish to maintain customers and discover the most important features affecting their loyalty, allowing organizations to make the right decisions at the right time to keep customers. The remainder of this paper is as follows: Sect. 2 discusses the literature reviewed for this work; Sect. 3 provides a description of the proposed techniques; Sect. 4 focuses on Empirical studies; Sect. 5 presents the results and discussion; Sect. 6 features further discussion highlights; and, finally, Sect. 7 presents the conclusion.

2   Related Work

Many studies have used various techniques for data mining to make churn predictions. This literature focuses on discussing related work that used data mining methods to apply a model of prediction.

In [3], the authors examined the problem of customer churn in banks and found that due to the intense competition between banks, they resorted to looking for intelligent ways to help them make decisions to win and maintain their customers. Consequently, the researchers proposed a model to predict customer churn using the neural network algorithm in the Alyuda NeuroInteligence software package. Using a database consisting of 1,866 customers from a small Croatian bank, they put forward the assumption that customers who use more than three products are likely to be loyal customers, whereas those who use less could pose a risk of leaving. One of the most important results they discovered was that there is the greatest risk of customer churn is among young people (students) who use less than three products.

In [4], the authors addressed the issue of customer churn. The main applied technique used data mining to extract useful data by developing the decision tree algorithm, which allows branch managers to identify who the most likely customers to leave are. They applied their study based on a dataset of random samples, consisting of 4,383 customers using electronic banking services. The technique highlights the characteristics of customers, thus enabling branch managers to introduce more marketing tools to retain them. However, the study could be more useful if it was supported by proven techniques that can be used to retain customers.

In [5], the authors discussed the problem of customer churn, which creates major issues for both enterprises and service providers, especially telecommunication companies. The research conducted a comparative study of churn prediction performance among two machine learning algorithms: K-nearest neighbor and decision tree. Their efficiency and performance were contrasted against the prediction issue churn and used performance evaluation criteria for the algorithms by precision, F-measure, recall, and accuracy. To reduce the noise and eliminate undesired information, filters processed the input data of each of the algorithms. The cleaned data were subsequently divided into training and testing sets. Training sets were then modeled to produce the desired output using algorithms. The algorithm for the K-nearest neighbor was set to the standard number of k = 5 neighbors. The standard Gini index criterion, which has a maximum depth of a size equal to 20, was used for the decision tree algorithm. There were 3,333 samples (customers) of 20 separate variables in the dataset that were used to compare all telecommunications firms' algorithms. It was found that the main outcomes were 92.6% more accurate than the K-nearest neighbor algorithm in terms of performance assessment criteria for the algorithms in the decision tree.

In [6], the authors discussed the churn that hinders the number of profitable customers from increasing, and it is the biggest challenge to sustain a telecommunication network. The key proposal included two models with a high degree of precision, and estimated customer churn. The logistic regression model was the first model, and its accuracy was improved by modifying it with the regularization parameter set to 0.01. The second model was the multilayer neural network (MLP). The dataset contained about four thousand lines. The main finding was 87.52% accurate using the logistic regression model, and it was 94.19% accurate using the neural network model. The two models' accuracy ratios were appropriate. However, some of the features were irrelevant in affecting the prediction results.

In [7], the authors considered the problem of rapid growth in Software as a Service (SaaS) companies, taking into account the existence of one source of income for the company, specifically the monthly fee for each customer. The customer behavior here needed to be analyzed to predict the factors that contribute to positive change and prevent customers from disconnection, which is essential and necessary. For companies to survive, they suggested building a predictive model using a time series perspective instead of using the traditional method of taking the variables accumulated from the moment the customer joined the company. They suggested four major algorithms: principal component analysis (PCA), logistic regression, random forest, and Extreme Gradient Boosting (XGBoost) trees using a traditional classification algorithm. They applied these algorithms to a set of data from multiple sources: the company's Microsoft SQL server, billing system, and Customer Relationship Management (CRM) platform, which contained 8,047 observations of 21 variables. The most important results came from the highest performance algorithms, namely, XGBoost and the logistic regression of 0.7257 and 0.7526, therefore they adopted the model XGBoost as the final model.

In [8], the authors explored customer churn in e-retail, building a model to make predictions using a data mining approach. They studied the problem in the context of a research community in North America of an E-retailer, investigating the characteristics of customers who reached 0.5 million in number. Furthermore, the attribute was classified into seven different categories: customer information, customer sales, demographic features, frequency, product sales, behavior, and experience. The number of all of the attributes was 35. Hadoop stack tools and classification were used with the following algorithms: logistic regression with L1 regularization, SVM, and gradient boost. The results revealed the significant impact of the following features: click/blogging, marketing campaign, customer behavior, and experience in customer momentum, to predict customers' churn. Besides, they found that the best prediction model was the GBM, with an accuracy of 75%.

In conclusion, previous studies have shown many differences in the field of application, such as the field of banking [3, 4], communications [5, 6], companies [7], and e-retailers [8]. However, their goal was the same, namely, to build a model to predict customer churn using a range of different algorithms. Of these algorithms, the ones that showed the best results were the decision tree algorithm, which outperformed the bank with a rate of 99.7%, followed by the neural network algorithm, with a rate of 94.19% in the field of communications. Therefore, these results motivated us to use these two algorithms in our study.

3 Description of Proposed Techniques

3.1 Decision Tree (J48)

The decision tree is one of the supervised classification algorithms and decision analysis tools. It uses a model in the form of a tree that includes potential outcomes. Each branch represents one of the options, and each one may also subdivide into other branches of future possibilities. It helps evaluate the choice between many options available and, in turn, make the best decision. It has four different types, which are: ID3, C45, C5.0, CART [9].

3.2 Multilayer Perceptron (MLP)

A Multilayer Perceptron (MLP) is a feed-forward artificial neural network model that contains one input layer, one output layer with at least one hidden layer. MLP utilizes a supervised learning technique called backpropagation for training. ANNs significance is to mainly solve three problems which are: classification, noise reduction, and extrapolation. Besides, MLP networks are one of the widest architectures with too many applications that can solve a lot of problems in various aspects of the knowledge areas; and the most significant areas are curve fitting, pattern recognition, process identification, and control time series forecasting, and system optimization [9, 10].

3.3 Random Forest (RF)

Random Forest is a machine learning algorithm that can be used for various tasks, including regression and classification [11]. It is a mixture of tree predictors, with each tree depending on independently sampled random vector values with the same distribution of all the trees in the forest. The random forest model combines predictions of estimators to produce a more accurate prediction. This algorithm requires two parameters: the ideal number of trees and the ideal depth of trees. The advantage RF presents variable significance estimates. They also give a superior technique for dealing with the missing data. The missing values are replaced by the variable that occurs most frequently in a specific node. RFs have the highest precision of all the classification methods available. With multiple variables running into thousands, the RF technique can also control big data. When a class is more infrequent than other data classes, it automatically balances information sets. In addition, the method also easily handles variables, making it ideal for complicated tasks.

4 Empirical Studies

4.1 Description of the Dataset

The data set contained 14 attributes and 10,000 instances and was extracted from a bank customers' data set using Kaggle, last updated was two years ago. The target variable was a binary variable reflecting whether the customer had left the bank (closed his/her account) or continued to be. Several pre-processes were followed to prepare the data, as shown in Fig. 1, where they were cleaned by deleting the unimportant attributes (customer number, customer name, and row number) and removing outliers. The dataset is imbalance; the class (No) the number of instances is 7,963, and the class (Yes) 2,037 instances. In this study, an over-sampling method using SMOTE (the technique of over-sampling of a synthetic minority), a widely used and available method of Weka as a supervised instance filter to address this problem. After the cleaning process, a transformation process was performed for some attributes. The target attribute (Exited) was converted from a numerical attribute to a categorical attribute, using an unsupervised filter (NumericToNominal). Lastly, the J48, RF, and neural network (MLP) algorithms were applied.

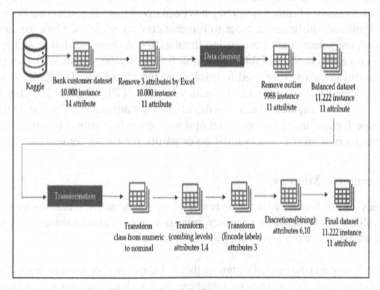

Fig. 1. Preparing dataset process

4.2 The Experimental Setup

Weka, a free software written under the JAVA General Public License that provides a collection of machine learning algorithms, was used in the study experiments [13]. Initially, the dataset was processed to prepare for the experiment. Three attributes were removed from the data using Excel, and then outliers were detected and removed using two filters (Interquartile Range and RemoveWith-Values). The issue of data imbalance was resolved in the quest to optimize classification, considering the cost of errors and overfitting issues, which can also lead to suboptimal results due to the high costs associated with misclassification of the minority class. The SMOTE algorithm (as supervised instance filter) in Weka was used to balance the data, which resulted in more synthetic instances for the minority class "1". After balancing the data, the total number of cases is 11222. The transformation step was critical to improving the model's performance and making the features easier to understand. The class value was transformed from a numeric value to a nominal value by WEKA using an unsupervised attribute filter (NumericToNominal). It was then combined with two attributes, and the labels for one attribute were encoded. Additionally, by binning the data, two numeric attributes were discretized; this step was completed using an unsupervised attribute filter (Discretize). Furthermore, the optimized parameters of J48, RF, and MLP were determined by setting the confidence factor of J48 and the minimum number of the object parameter. The RF number of the tree and the seed parameter were reset. In addition to adjusting MLP seed, learning rate, and the hidden layers parameter. As a consequence, the J48 performed higher, with an accuracy of 83.55% compared to 82.39% for the MLP, and the RF performed at 81.87%. Moreover, correlation coefficients and information gain between the class variable and each attribute were calculated in order to rank the attributes of the selected features. The findings are shown in Tables 3 and 4. Following that, the ability to enhance the accuracy of rating efficiency has been investigated by selecting features and defining significant attributes in order to forecast consumer churn. The classifiers were evaluated on the selected attributes using the the10-fold cross-validation and optimum parameters for each classifier. Table 5 and Table 6 illustrate the results. With these results, different partition ratios were used to implement the classifiers. The highest accuracy for J48, MLP, and RF was found to be achieved with a (70:30) ratio, 70% for training data and 30% for testing data. Table 7 summarizes the findings. Finally, the final studies have been performed using the optimum options for the best subset features to achieve the best results in terms of cross-validation or partition ratio outcome.

4.3 Performance Measures

In this study, to produce more accurate results, four performance measures were considered to evaluate each classifier: accuracy, precision, recall, and f-measure. All of these measures are based on the following possibilities:

- TP: True Positive is the total of instances that a churn customer was correctly classified.
- FP: False positive is the total of instances that a churn customer was incorrectly classified.

- TN: True Negative is the total of instances that a non_churn customer was correctly classified.
- FN: False negative is the total of instances that a non_churn customer was incorrectly classified.

$$Accuracy = \frac{TP + TN}{TP + TN + FN + FP} \quad (1) \qquad Recal = \frac{TP}{TP + FN} \quad (3)$$

$$Precision = \frac{TP}{TP + FP} \quad (2) \qquad f-meas = \frac{2 \times precision \times recall}{precision \times recall} \quad (4)$$

5 Strategy of Optimization

In seeking to improve the classification results, the Weka meta-learner (CV Parameter Selection) search methodology was used to obtain better performance based on accuracy [14]. Table 1 shows the default and optimum parameters for each classifier. Table 2 shows the comparison between the findings associated with optimal parameters and that of their default counterparts. It can be noticed that the algorithms accuracy increases when the optimum parameters are used in comparison to the default parameters.

Table 1. Classifiers parameters: default and optimal

Model	Values of parameters		
	Parameters	Default	Optimal
J48	Confidence factor	0.25	0.2
	Minimum number of objects	2	4
MLP	Seed	0	0
	Hidden layers	A	a
	Learning rate	0.3	0.116
RF	NumIterations (number of trees)	100	50
	Seed	1	32

Table 2. Classifiers performance: default and optimal parameters

Model	Performance (Accuracy)	
	Default value	Optimal value
J48	83.34%	83.55%
MLP	80.46%	82.39%
RF	81.66%	81.87%

6 Results and Discussion

6.1 The Impact of Features Selection on the Dataset

The Information Gain (InfoGain) and correlation-based features selection method were used to select the best performing subset, along with the most significant attributes with the highest impact on prediction of customer churn. The correlation coefficient was used to rank the attributes based on the Pearson values, from highest to the lowest variable relationship with the class variable (output), as shown in Table 3. In addition, InfoGain was applied to classify the features based on the class information gain measure, as shown in Table 4. The backward selection method begins with building a model with all of the features required in their order of significance. The least important feature is removed. Then the next model is built using the remaining features. Features are continually removed, and models continue to be built until one feature remains. The results of the Info Gain and correlation-based features selection method are presented in Table 5 and Table 6. It was observed that the best performance required the use of all of the features (10 features). Furthermore, accuracy diminished when the number of factors was reduced.

Table 3. The correlation of each attribute and the target

Table 4. The information gain of each attribute and the target

Rank	Attributes	Correlation	Rank	Attributes	InfoGain
1	Gender	0.1858	1	NumOfProducts	0.2075
2	IsActiveMember	0.1804	2	Tenure	0.1813
3	Geography	0.1314	3	IsActiveMember	0.1256
4	Balance	0.0812	4	HasCrCard	0.0862
5	NumOfProducts	0.0586	5	Geography	0.0332
6	CreditScore	0.0197	6	Gender	0.0251
7	Tenure	0.0137	7	Balance	0.0185
8	Age	0.0102	8	Age	0.0026
9	EstimatedSalary	0.0070	9	CreditScore	0.0014
10	HasCrCard	0.0003	10	EstimatedSalary	0.0004

6.2 The Impact of Different Partition Ratios on the Dataset

After identifying the best features, it became clear that all of the features are important in both InfoGain and the correlation-based features selection method. The performance of each classifier was evaluated by performing many experiments on the data using different partition ratios ranging from 50 to 80. The results of the direct partition of each classifier are shown in Table 7.

Table 5. Correlation-based feature selection results

Number of features	J48	MLP	RF	AVG (%)
10	83.55	82.39	81.87	82.60
9	83.38	81.78	81.09	82.08
8	83.42	81.63	80.14	81.73
7	82.97	81.58	79.66	81.41
6	83.07	81.92	82.41	82.47
5	83.24	80.86	83.57	82.56
4	77.78	77.13	78.08	77.66
3	77.07	76.43	77.06	76.85
2	76.20	70.75	76.20	74.38
1	70.95	70.95	70.95	70.95

Table 6. InfoGain feature selection results

Number of features	J48	MLP	RF	AVG (%)
10	83.55	82.39	81.87	82.60
9	83.51	82.48	80.40	82.13
8	83.50	81.51	81.25	82.09
7	83.09	80.93	81.17	81.73
6	81.90	81.11	82.57	81.86
5	82.22	80.25	83.34	81.94
4	81.63	76.84	83.27	80.58
3	80.71	75.36	83.27	79.78
2	80.81	77.31	83.19	80.44
1	78.19	70.95	78.20	73.37

Table 7. Results of different partition ratios

Partition ratio	The performance			
	J48	MLP	RF	AVG
50:50	83.1937%	80.0036%	78.2392%	80.4788%
60:40	83.2479%	80.2629%	79.1045%	80.8717%
70:30	83.4868%	80.1901%	79.4179%	81.0316%
80:20	82.3975%	80.8378%	77.6292%	80.2881%

6.3 The Comparison Between 10-Fold Cross Validation with Direct Partition Techniques

When comparing the two validation methods (10-fold cross-validation and direct partition ratio), all of the classifiers that used 10-fold cross-validation gained a higher accuracy, as shown in Table 8.

Table 8. 10-fold cross validation comparison with Direct Partition techniques

Techniques	Proposed model			
	J48	MLP	RF	AVG
10-fold validation	83.5502%	82.3917%	81.8749%	82.6056%
Partition ratio	83.4868%	80.1901%	79.4179%	81.0316%

7　Further Discussion

The final customer churn prediction model was built using all of the features with the optimum parameters achieved, as noted in Table 9. Using the 10-fold cross-validation method, the researchers of this study were able to produce perfect results for each of the following classifiers: RF, J48, and MLP. The J48 outperformed MLP and RF in predicting customer churn with an accuracy of 83.0957%. The classification performance was increased by choosing all the features (10 features) using the optimal criteria for each classifier, as shown in Fig. 2. Through this process, we were able to identify important features that had a significant impact on the ability to predict customer churn in the banking sector, specifically: Credit Score, Geography, Gender, Age, Tenure, Balance, Number of products, Has Credit Card, Is an Active Member, and Estimated Salary. Identifying essential features can help prevent customers from leaving and f competition from rival banks. The Receiver Operating Characteristic (ROC) curve is another indicator of how the model of classification performs. The proximity and placement of this curve to the left-hand side (on the top) indicate that the experiment's accuracy is high. Overall, the area under the curve for each classifier as noted in Fig. 3, shows that the most suitable classifier is determined to be J48 compared to MLP and RF. According to table 9, J48 performed higher compared to the rest of the algorithms across all of the measures. Confusion matrices offer another look at actual and predicted classes by J48, MLP, and RF in Tables 10, 11, 12, respectively. The most important measure to check in the confusion matrices below is the rate of False Negative (FN), as an increase in its rate adversely affects the bank's profits. The lowest FN rate was using RF, J48, and MLP, respectively.

Table 9. The performance of proposed model.

Proposed model	J48				RF				MLP			
Performance measures	ACC%	PREC	REC	F-M	ACC%	PREC	REC	F-M	ACC%	PREC	REC	F-M
Results of final model	83.55	0.578	0.826	0.836	81.87	0.813	0.819	0.813	82.39	0.548	0.815	0.824

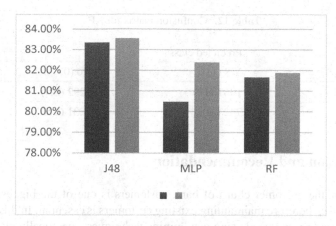

Fig. 2. The result of default and optimized model.

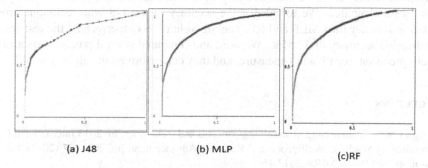

(a) J48   (b) MLP

(c)RF

Fig. 3. The ROC curve for Yes (1) class for each classifier.

Table 10. Confusion matrix for J48

Actual class	Predicted class	
	Yes (1)	No (0)
Yes (1)	1843 (TP)	1416 (FN)
No (0)	430 (FP)	7533 (TN)

Table 11. Confusion matrix for MLP

Actual class	Predicted class	
	Yes (1)	No (0)
Yes (1)	1413 (TP)	1846 (FN)
No (0)	7400 (FP)	563 (TN)

Table 12. Confusion matrix for RF

Actual class	Predicted class	
	Yes (1)	No (0)
Yes (1)	1967 (TP)	1292 (FN)
No (0)	742 (FP)	7221 (TN)

8 Conclusion and Recommendation

In conclusion, the customer churn of bank customers is one of the biggest causes of losses for banks, therefore maintaining existing customers is essential. In this study, one bank's database was analyzed using data mining techniques, specifically classification. Three algorithms were also applied to classify decision tree J48, MLP, and RF. Two techniques, namely cross-validation, and the partition ratio were applied and then their results were compared. We found that the accuracy of the J48 with technique cross-validation is better than MLP and RF. The algorithm J48 outperformed the rest with a classification accuracy of 83.55%. We used and compared several precision measures, namely precision, recall, and f-measure, and they outperformed the algorithm J48.

References

1. Agarwal, S.: Data mining: data mining concepts and techniques. In: 2013 International Conference on Machine Intelligence and Research Advancement, pp. 203–207 (2013). https://doi.org/10.1109/ICMIRA.2013.45
2. Dai, T.: International trade e-commerce based on data mining. In: Proceedings - 2014 IEEE Workshop on Advanced Research and Technology in Industry Applications, WARTIA 2014, pp. 703–705 (2014). https://doi.org/10.1109/WARTIA.2014.6976362
3. Bilal Zoric, A.: Predicting customer churn in banking industry using neural networks. Interdiscip. Descr. Complex Syst. **14**(2), 116–124 (2016). https://doi.org/10.7906/indecs.14.2.1
4. Keramati, A., Ghaneei, H., Mirmohammadi, S.M.: Developing a prediction model for customer churn from electronic banking services using data mining. Financ. Innov. **2**(1), 1–13 (2016). https://doi.org/10.1186/s40854-016-0029-6
5. Lazarov, V., Capota, M.: Churn prediction: a comparative study using KNN and decision trees. In: 2019 Sixth HCT Information Technology Trends, no. 1, pp. 182–186 (2019)
6. Kim, K., Lee, J.H.: Bayesian optimization of customer churn predictive model. In: Joint 10th International Conference on Soft Computing and Intelligent Systems and 19th International Symposium on Advanced Intelligent System, pp. 85–88 (2018). https://doi.org/10.1109/SCIS-ISIS.2018.00024
7. Ge, Y., He, S., Xiong, J., Brown, D.E.: Customer churn analysis for a software-as-a-service company. In: 2017 Systems and Information Engineering Design Symposium, SIEDS 2017, pp. 106–111 (2017). https://doi.org/10.1109/SIEDS.2017.7937698
8. Subramanya, K.B., Somani, A.K.: Enhanced feature mining and classifier models to predict customer churn for an E-retailer. Big Data Anal. Tools Technol. Eff. Plan., 293–309 (2017). https://doi.org/10.1201/b21822

9. Han, J., Kamber, M., Pei, J.: Data Mining Concepts and Techniques, 3rd edn. Morgan Kaufmann Publishers Inc., San Francisco (2011)
10. López-Gil, J.M., Virgili-Gomá, J., Gil, R., García, R.: Method for improving EEG based emotion recognition by combining it with synchronized biometric and eye tracking technologies in a non-invasive and low cost way. Front. Comput. Neurosci. **10** (2016). https://doi.org/10.3389/fncom.2016.00085
11. Pavlov, Y.L.: Random forests. In: Random Forest, pp. 1–122 (2019). https://doi.org/10.1201/9780429469275-8
12. What Is a Good Credit Score? - Experian. https://www.experian.com/blogs/ask-experian/credit-education/score-basics/what-is-a-good-credit-score/. Accessed 08 Jan 2021
13. Daw, S., Basak, R.: Machine learning applications using Waikato environment for knowledge analysis. In: Proceedings of the 4th Fourth International Conference on Computing Methodologies and Communication ICCMC 2020, pp. 346–351, February 2020. https://doi.org/10.1109/ICCMC48092.2020.ICCMC-00065
14. Tufi, D.: Proceedings of the 10th International Conference 'Linguistic Resources and Tools for Processing the Romanian Language' 18–19 September 2014, September 2014

Early Prediction of Diabetes Disease Based on Data Mining Techniques

Salma N. Elsadek⑩, Lama S. Alshehri, Rawan A. Alqhatani, Zainah A. Algarni,
Linda O. Elbadry⑩, and Eyman A. Alyahyan^(✉)⑩

Computer Science Department, College of Science and Humanities, Imam Abdulrahman Bin
Faisal University, P.O. Box 31961, Jubail, Kingdom of Saudi Arabia
eaalyahyan@iau.edu.sa

Abstract. As a common yet chronic disease, Diabetes Mellitus (DM) affects millions of people all over the world. Some groups are more vulnerable to diabetes in comparison to others, such as people with a family history, and those suffering from obesity. Early detection of such people, in conjunction with preventive measures, can go a long way in saving the lives of people who are most likely to be infected with these diseases and avoid suffering. Consequently, Prediction solutions have been found using data mining techniques which help to discover hidden information about the disease and supports decision-making. This study aims to ensure that diabetes is predicated at the initial stages using two algorithms of machine learning: Random Forest and Multi-Layer Perceptron (MLP) using the WEKA environment to estimate the accuracy. The experiments were applied using a dataset obtained from the Machine Learning Repository of UCI. The dataset comprises 16 attributes and 520 instances collected via questionnaires from patients at Sylhet-based Sylhet Diabetes Hospital (Bangladesh). According to the findings, The RF method has been found to yield better results than MLP when it comes to enabling early prediction of diabetes with a high accuracy of 97.88% .

Keywords: Data mining · Diabetes · Random Forest · MLP

1 Introduction

Diabetes mellitus is one of the most chronic and frequent diseases, affecting422 million people globally; by 2030, over 552 million people around the world are likely to experience diabetes [1]. World Health Organization (WHO), suggests that1.6 million people die due to diabetes annually. It is caused when the pancreas have trouble producing and building glucose in the blood, which leads to an increase in blood sugar concentration [2]. Glucose, which is blood sugar, comes from different types of foods. Insulin t helps the glucose get into the body cells to give them energy. Diabetes is of three types, the most dangerous of which is type1, and can cause serious issues. For example, it can damage kidneys, eyes, and other organs. Type 2, and Gestational Diabetes are not as dangerous as type 1, and can be controlled by organizing meals and making exercises.

© IFIP International Federation for Information Processing 2021
Published by Springer Nature Switzerland AG 2021
V. Krishnamurthy et al. (Eds.): ICCIDS 2021, IFIP AICT 611, pp. 40–51, 2021.
https://doi.org/10.1007/978-3-030-92600-7_4

Diabetes may lead to serious health complications or in extreme cases, lead to death if it is not discovered in a timely manner [3].

Diabetes is the biggest and most important epidemic confronting global economy. Despite allocating large financial and medical resources to confront it in developed countries, decisive economic considerations have not been reached to arrest or reduce its costly complications that are now increasingly affecting the quality of healthcare. For example the United States of America spent 327 billion dollar in diagnosing diabetes in 2017, which includes 237 billion in the form of direct medical costs and reduced productivity worth 90 billion [4] This requires us to adopt a long-term outlook to reduce diabetes and lower the economic cost, while ensuring early diagnosis and attention to avoid complications such as kidney failure, retinopathy, loss of limbs, arterial stiffness, stroke and heart disease are the most important means to reduce the economic burden of diabetes.

The process of finding meaningful knowledge from hidden patterns is called data mining, which assumes great significance in healthcare, especially in forecasting diseases like diabetes and cancer. It can also facilitate the diagnosis for doctors in making their clinical decisions [5] by extracting knowledge from copious amounts of data to identify patterns as well as to establish relationships that help solve problems. Applying data mining techniques will facilitate the early prediction of diabetes which improves treatment [6]. Then it can be used to create an efficient decision-making process in the medical field. Therefore, the accurate right decision planning of predictive data mining is a highly creative methodology that the (WHO) would do well to take seriously [7].

In the majority of studies conducted so far, machine learning algorithms have been used more than deep learning. The Pima Indians Diabetes Database (PIDD) was the most widely used. Different algorithms were applied to this database, therefore, the result differed from one study to another regarding the best algorithm to apply. In our research, we used a data set recently collected from the patients of Sylhet (Bangladesh)-based Diabetes Hospital. Thus far, only two similar studies have been carried out using the same dataset [8, 9]. In [8], the authors applied Naive Bayes (NB), J48 Decision Trees, Logistic Regression, and Random Forest (RF), while in [9], the authors investigated the disease on the basis of Naïve Bayes, Random Tree, SVM, K-NN, Bayes Network, J48, and Random Forest.

This study aims to build two data mining models using classification methods Random Forest (RF) and Multilayer Perceptron (MLP) using WEKA. Since no study thus far on the Sylhet dataset has compared MLP with RF, we're seeking a comparison between them to determine the best performance to help forecast the onset of diabetes using different measurements: accuracy and Roc curve. Also, MLP and Random Forest are essential data mining methods, and they can diagnose diseases at an early stage and obtain high performance compared with other classifiers. Thus, these methods are adopted in this study. In addition, this study attempts to determine the most salient issues (predictive) that can impact diabetes; the findings can also be used to issue early warnings about the disease.

The organization of this paper is done in the following manner. Sect. 2 reviews associated literature work. Sect. 3 contains a description of the techniques proposed in this study. Sect. 4 contains dataset description, experimental setup, whereas Sect. 5 elaborates on Optimization Strategy. Results and discussion is presented in Sect. 6 whereas Sect. 7 contains the conclusion.

2 Related Work

In [8], the authors aimed to discover the best algorithm for forecasting the risk factors for diagnosis. The dataset comprises 520 instances as well as 16 attributes gathered via Sylhet Diabetes Hospital. The dataset was assessed using J48 Decision Trees, RF and Logistic Regression (LR) algorithms. As per the findings, the accuracy of RF was the highest at 97.4% as compared to others.

In [9], the researchers proposed a classifier that classifies diabetes using data mining techniques. They study seven algorithms: Naïve Bayes, Random Tree, SVM, K-NN, Bayes Network, J48, and Random Forest. This study attempted to assess the algorithms' execution for a diabetes dataset using the WEKA program and applied to Sylhet Diabetes Hospital. As per this study, the highest accuracy was attained by k-NN (98.07%), and also helped classify diabetes within the dataset.

In [10], the authors discussed a method for determining the incidence of diabetes using machine learning algorithms. This paper aimed to discover diabetes in its early stages and design a model that can accurately predict the prospect of getting diagnosed with this disease. Three machine learning algorithms DT, SVM, and NB were utilized and the evaluation of the algorithm took place based on accuracy, F-scaling, and other measures, and to determine the correct classification of cases. The results showed that the NB algorithm achieved higher accuracy by 76.30%.

In [11], the authors aimed to early prediction of diabetes using various techniques of data mining. The dataset comprised 768 instances and the dataset was assessed using WEKA and MATLAB tool with Naive Bayes, MLP, Bayesian Network, PLS-LDA, Homogeneity-Based, ANN, C4.5, Amalgam KNN, ANFIS, and Modified J48. The results showed that Modified J48 had the highest accuracy of 99.87% for predicting the disease, while the ANN algorithm had the worst accuracy of 73.44%.

In [12], the authors presented a novel model to forecast type 2 diabetes using the techniques mentioned above to not only enhance the prediction model's accuracy, but also to increase its suitability to more than a single dataset. The main technique applied was the enhanced algorithm of LR and K-means cluster. According to the findings, the accuracy of this model 3.04% higher as compared to other findings, which revealed 95.42%. Also, evaluating the model using two datasets, the model revealed an accuracy of 90.7% and 94, respectively, providing that the performance of the model was good.

In [13], the problem of early prediction of gestational diabetes was discussed. The authors aimed to compare three data mining algorithms depending on some attributes to provide the best algorithm. The main technique applied to the dataset was NB, KNN, and DT. The results showed that the DT's accuracy was the highest with 75.65%, while the accuracy of KNN was the lowest (65.16%).

In [14], the authors studied the same problems. The dataset used in this research comprised nine attributes with 768 instances. The relationship between the attributes was identified by the researchers. The main techniques applied in this dataset included RF, ANN, as well as K-means clustering. The accuracy of ANN was the highest at 75.7%. The results also revealed a tight linkage between diabetes, glucose and Body Mass Index (BMI).

In [15], the authors have aimed to build a model of prediction for three complications related to diabetes in Indonesia and to ascertain their relationship to characteristics, including age and blood glucose level. The database was collected from three sources, consists of 158 medical records. Three data mining techniques used included C4.5 decision tree, NB, and k-mean clustering for data analysis purposes. The authors assessed performance through clustering and classification techniques. The classification technique gives better information and performance compared to the clustering technique. The researchers used accuracy as a metric for evaluating this model's accuracy. The accuracy of Diabetes Complication Prediction Model was revealed to be 68%. (Table 1) summarizes the previous literature in the order in which it is presented above.

Table 1. Related work

Ref	Year	Proposed method	Dataset	Best method
8	2020	NB, J48 D, Logistic Regression, and Random Forest	520 instances from Diabetes symptom Dataset	Random Forest 97.4%
9	2021	NB, Random Tree, SVM, k-NN, Bayes Network, J48, and Random Forest	520 instances from Diabetes symptom Dataset	k-NN 98.07%
10	2018	DT, SVM and NB	768 instances from Pima Indians Diabetes Database (PIDD)	NB 76.30%
11	2018	NB, MLP, Bayesian Network, PLS-LDA, Homogeneity-Based, ANN, C4.5, Amalgam KNN, ANFIS, and Modified J48	768 instances from PIMA Indian Dataset	Modified J48 99.87%
12	2018	The improved K-means and Logistic regression algorithm	768 instances from Pima Indians Diabetes Database (PIDD)	Proposed model 95.42%
13	2018	DT, NB, and K-NN	768 instances from Pima Indians Diabetes Database (PIDD)	DT 75.65%

(*continued*)

Table 1. (*continued*)

Ref	Year	Proposed method	Dataset	Best method
14	2019	ANN, Random Forest, and K-means	768 instances from the National Institute of Diabetes and Digestive and Kidney Diseases	ANN 75.7%
15	2019	NB Tree, C4.5 and k means	158 instances from three sources (Sri Pamela Hospital and Kumpulan Pane Hospital)	Proposed model 68%

3   Description of the Proposed Techniques

3.1   ANN

ANN or Artificial Neural Network is a class of algorithms that uses artificial intelligence technology. ANN is an interconnected set of virtual neurons created by computer programs that aims to collect knowledge through training and then storing it via the neurons' connecting forces referred to as synaptic weights [16]. ANN aims to simulate the intelligence of the human brain. The idea of neural networks goes back to two researchers at the University of Chicago, Warren McCullough, a neuroscientist, and Walter Bates, a mathematician, in 1943 [17].

The essential elements in ANN are nodes called neurons and connections between them. As mentioned before, the training process determines the value of connection weights and is paramount to discern its learning ability. The three layers of ANN are Output, Hidden, and Input [18] Multilayer Perceptron (MLP) is one of ANN algorithms, characterized by adding hidden layers to simple perception [19]. Figure 1 shows a schematic graph of MLP [20].

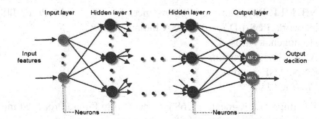

Fig. 1. A schematic view of an MLP neural network [20]

3.2 Random Forest

This form of machine learning technology is an integrated and multi-purpose tool that uses a set of data and variables to build more than one decision tree. It is then combined to obtain more stable and accurate forecasts. RF can be expedited on large datasets and is used in regression as well as classification problems. Leo Breiman had proposed RF in 2001 [21, 22]. RF needs two parameters to be tuned: total number of variables and trees [23]. Figure 2 shows a pictorial representation of RF [22].

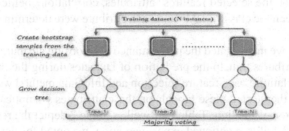

Fig. 2. Pictorial representation of RF [22]

4   Contains Dataset Description and Setup

4.1 Description of the Dataset

The dataset was extracted from the UCI repository, and collected using a questionnaire shared with the patients of Sylhet Diabetes Hospital. It contains 17 attributes, and 520 instances, the attributes are all nominal except age which is numeric. The range of age of the participants was 20-65 and above. Figure 3 summarizes the data preparation's approach.

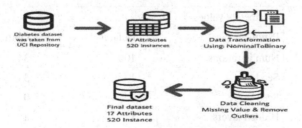

Fig. 3. Process of dataset preparation

4.2 Experimental Setup

The experiment was applied using WEKA, WEKA is a workspace that provides an excellent mechanism to examine common problems of data mining like clustering, classification and it facilitates comparison of different machine learning techniques [24].

Initially, the data set was preprocessed in preparation for the experiment. The nom inal data were transformed to numeric binary data using a supervised attribute filter (NominalToBinary). No missing data was observed. Also, an unsupervised attribute filter (InterquartileRange) was used to handle outliers in the data set in WEKA.

Additionally, the optimized parameters of RF, and MLP were determined by adjusting the number of the tree, and the seed parameter was resetting. In addition to adjusting the seed, the learning rate, and the hidden layers parameter for the MLP. As per the findings, the accuracy of RF and MLF was 97.88% and 96.34%, respectively.

Besides, to rank the selected features' attributes, correlation coefficients and information gain between the class variable and each attribute were determined. Tables 5 and 6 show the results.

Subsequently, we investigated the performance by choosing attributes and determining significant attributes to help the prediction of Diabetes during the infancy stage. In this process, correlation-based feature selection and InfoGain method were utilized. For all classifiers, we then tested these models on these attributes via optical parameters as well as 10-fold cross-validation. Table 6 as well as Table 7 depict the results.

Finally, from the aforementioned experimentation, we opted for the optimal choice encompassing all attributes for undertaking the model's development and attaining the optimal outcome in terms of cross-validation.

5   Strategy of Optimization

Intending to optimize the results of classification, we used the CVParameterSelection, a search methodology for undertaking optimal parameters for each classifier using WEKA [25]. Our attempt was to yield the best criteria of performance based on accuracy using 10 cross-validations. Table 2 depicts both default parameters as well as optimal parameters.

Table 2. Parameters for all classifiers: default and optimal

Model	Parameters values		
	Parameters	Default	Optimal
RF	Num iterations	100	34
	Seed	1	34
MLP	Seed	0	0
	Hidden layers	a	a
	Learning rate	0.3	0.3

Table 3 shows the comparison of findings associated with optimal parameters and that of their default counterparts. It can be seen that the RF algorithm's accuracy increases, but that of MLP remains unchanged.

Table 3. Classifiers performance: default and optimal parameters

Model	Accuracy of performance	
	Default	Optimal
RF	97.11%	97.88%
MLP	96.34%	96.34%

6   Results and Discussion

6.1   Effect of Feature Selection on the Dataset

The information gain and correlation-based selection of feature methods were used to determine the best performing subset along with the most salient attributes to ensure early prediction of Diabetes. The correlation coefficient was used to rank the features based on the Pearson values, from the highest to the lowest variable relationship with the class variable (output), as shown in Table 4. Besides, InfoGain was applied to classify the features based on the class information gain measure, as shown in Table 5.

Table 6 and Table 7 clearly point out that the implementation of all features (16 in totality) helped obtain the best performance. Moreover, the results have been shown that decreasing the number of features leads to a decrease in the percent-age of accuracy.

6.2   Further Discussion

Table 8 and Fig. 4 show the model that was used to ensure early prediction of diabetes using 16 features involving optimal parameters. The technique of 10-fold cross-validation was used to attain optimal findings of MLP and RF. RF was found to outperform MLP in terms of forecasting Diabetes. Its accuracy was found to be 97.88%. Meanwhile, the accuracy was negatively impacted when the number of features was lowered and enhanced after using all features. Based on these findings, it can be inferred that it is necessary to incorporate all features to increase the accuracy of forecasting diabetes at an early stage. The Receiver Operating Characteristic (ROC) curve is another indicator of how the model of classification performs. Figures 5 and 6 illustrate that the proximity and placement of this curve to the left-hand side (on the top) indicate that the experiment's accuracy is high. Overall, the area under the curve for each classifier shows that the most suitable classifier is determined to be RF compared to MLP.

Table 4. The correlation of each attribute and the target

Sequence	Attribute name	Correlation
1	Polyuria	0.6659
2	Polydipsia	0.6487
3	Gender	0.4492
4	Sudden weight loss	0.4366
5	Partial paresis	0.4323
6	Polyphagia	0.3425
7	Irritability	0.2995
8	Alopecia	0.2675
9	Visual blurring	0.2513
10	Weakness	0.2433
11	Muscle stiffness	0.1225
12	Genital thrush	0.1103
13	Age	0.1087
14	Obesity	0.0722
15	Delayed healing	0.047
16	Itching	0.0134

Table 5. The information gain of each attribute and the target

Sequence	Attribute name	InfoGain
1	Polyuria	0.36225
2	Polydipsia	0.35906
3	Gender	0.16342
4	Sudden weight loss	0.14877
5	Partial paresis	0.14465
6	Polyphagia	0.14465
7	Irritability	0.07287
8	Alopecia	0.05116
9	Visual blurring	0.04661
10	Weakness	0.04267
11	Age	0.02239
12	Muscle stiffness	0.01097
13	Genital thrush	0.00905
14	Itching	0
15	Delayed healing	0
16	Obesity	0

Table 6. InfoGain feature selection results

Number of features	Features	RF	MLP	AVG
Using features: all (16)	All	97.88%	96.34%	97.09%
Using 8 features	Polydipsia, Polyuria, Sex Sudden loss of weight, Partial paresis, Irritability, Alopecia, Polyphagia	94.03%	93.65%	93.94%
Using 4 features	Polydipsia, Polyuria, Sex sudeen loss of weight	88.84%	89.42%	89.13%
Using 2 features	Polyuria, Polydipsia	86.92%	86.92%	86.92%

Table 7. Correlation-based feature selection results

Number of features	Features	RF	MLP	AVG
Using features: all (16)	All	97.88%	96.34%	97.09%
Using 8 features	Polydipsia, Polyuria, Sex Sudeen loss of weight, Partial paresis, Alopecia, Irritability, Polyphagia	94.03%	93.65%	93.94%
Using 4 features	Polydipsia, Polyuria, Sex sudeen loss of weight	88.84%	89.42%	89.13%
Using 2 features	Polyuria, Polydipsia	86.92%	86.92%	86.92%

Table 8. Findings of best options based on optimal parameters

Techniques	RF	MLP
10-fold validation	97.88%	96.34%

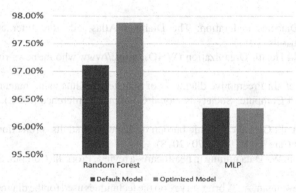

Fig. 4. Default and optimized models: a comparison

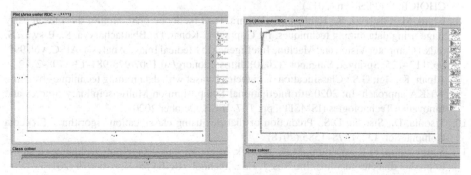

Fig. 5. Positive class RF ROC curve   **Fig. 6.** Positive class MLP ROC curve

7 Conclusion

Diabetes is characterized by a high blood sugar level due to the inability of the pancreas to generate sufficient insulin. It can also be caused when the cells of the human body are unable to respond adequately to the insulin produced in the body. It prevents the body from ensuring proper utilization of energy. Early prediction of diabetes can go a long way in helping people avert the negative economic consequences that emerge from the constant rise in the number of cases involving this serious disease that affects people from all over the world. In this context, data mining has emerged as an important tool that facilitates the discovery of hidden information. This study implemented two classifiers, random forest and MLP, in order to obtain the best accuracy rates for estimating diabetes. Accordingly, the random forest method was found to yield better results than MLP when it comes to enabling early prediction of diabetes with high accuracy of 97.88%. The experiment revealed that both algorithms work better when all features are utilized; therefore, each of these features was used to achieve the highest possible accuracy ratio. In the aftermath of the optimization strategy, the random forest was also found to outperform the results derived in [8], which was found to have an accuracy of 97.4%.

References

1. International Diabetes Federation: The Diabetes Atlas, 5th edn. International Diabetes Federation, Brussels (2011)
2. Diabetes, World Health Organization (WHO). https://www.who.int/news-room/fact-sheets/detail/diabetes
3. Alassaf, R.A., et al.: Preemptive diagnosis of diabetes mellitus using machine learning. In: 2018 21st Saudi Computer Society National Computer Conference (NCC), pp. 1–5. IEEE, April 2018
4. Bommer, C., et al.: Global economic burden of diabetes in adults: projections from 2015 to 2030. Diabetes Care **41**(5), 963–970 (2018)
5. Jothi, N., Husain, W.: Data mining in healthcare–a review. Procedia Comput. Sci. **72**, 306–313 (2015)
6. Agrawal, P., Dewangan, A.: A brief survey on the techniques used for the diagnosis of diabetes-mellitus. Int. Res. J. Eng. Tech. IRJET **2**, 1039–1043 (2015)
7. Cheung, J.Y.: Data Mining: Concepts and Techniques. American Library Association DBA CHOICE, Middletown (2012)
8. Islam, M., Ferdousi, R., Rahman, S., Bushra, H.: Likelihood prediction of diabetes at early stage using data mining techniques. In: Gupta, M., Konar, D., Bhattacharyya, S., Biswas, S. (eds.) Computer Vision and Machine Intelligence in Medical Image Analysis. AISC, vol. 992, pp. 113–125. Springer, Singapore (2020). https://doi.org/10.1007/978-981-13-8798-2_12
9. Alpan, K., İlgi, G.S.: Classification of diabetes dataset with data mining techniques by using WEKA approach. In: 2020 4th International Symposium on Multidisciplinary Studies and Innovative Technologies (ISMSIT), pp. 1–7. IEEE, October 2020
10. Sisodia, D., Sisodia, D.S.: Prediction of diabetes using classification algorithms. Procedia Comput. Sci. **132**, 1578–1585 (2018)
11. Sengamuthu, M.R., Abirami, M.R., Karthik, M.D.: Various data mining techniques analysis to predict diabetes mellitus. Int. Res. J. Eng. Technol. (IRJET) **5**(5), 676–679 (2018)
12. Wu, H., Yang, S., Huang, Z., He, J., Wang, X.: Type 2 diabetes mellitus prediction model based on data mining. Inform. Med. Unlocked **10**, 100–107 (2018)

13. Azrar, A., Ali, Y., Awais, M., Zaheer, K.: Data mining models comparison for diabetes prediction. Int. J. Adv. Comput. Sci. Appl. (IJACSA) **9**, 320–323 (2018)·
14. Alam, T.M., et al.: A model for early prediction of diabetes. Inform. Med. Unlocked **16**, 100204 (2019)
15. Fiarni, C., Sipayung, E.M., Maemunah, S.: Analysis and prediction of diabetes complication disease using data mining algorithm. Procedia Comput. Sci. **161**, 44 (2019)
16. Jain, A., Kumar, A.: An evaluation of artificial neural network technique for the determination of infiltration model parameters. Appl. Soft Comput. **6**(3), 272–282 (2006)
17. Yaqub, F.: A Study on Artificial Neural Network
18. Grossi, E., Buscema, M.: Introduction to artificial neural networks. Eur. J. Gastroenterol. Hepatol. **19**(12), 1046–1054 (2007)
19. Azeez, O.S., Pradhan, B., Shafri, H.Z., Shukla, N., Lee, C.W., Rizeei, H.M.: Modeling of CO emissions from traffic vehicles using artificial neural networks. Appl. Sci. **9**(2), 313 (2019)
20. Ebrahimi, E., Mollazade, K., Arefi, A.: An expert system for classification of potato tubers using image processing and artificial neural networks. Int. J. Food Eng. **8**(4) (2012)
21. Breiman, L.: Random forests. Mach. Learn. **45**(1), 5–32 (2001)
22. Al Amrani, Y., Lazaar, M., El Kadiri, K.E.: Random forest and support vector machine based hybrid approach to sentiment analysis. Procedia Comput. Sci. **127**, 511–520 (2018)
23. Naghibi, S.A., Ahmadi, K., Daneshi, A.: Application of support vector machine, random forest, and genetic algorithm optimized random forest models in groundwater potential mapping. Water Resour. Manage **31**(9), 2761–2775 (2017)
24. Frank, E., Hall, M., Trigg, L., Holmes, G., Witten, I.H.: Data mining in bioinformatics using Weka. Bioinformatics **20**(15), 2479–2481 (2004)
25. https://weka.sourceforge.io/doc.dev/weka/classifiers/meta/CVParameterSelection.html.

An Application-Driven IoT Based Rooftop Farming Model for Urban Agriculture

Arjun Paramarthalingam[1]([envelope]) [ID], Amirthasaravanan Arivunambi[1] [ID],
and Sreedhar Thapasimony[2]

[1] CSE, University College of Engineering Villupuram, Villupuram, Tamil Nadu, India
[2] EEE, Rohini College of Engineering and Technology, Anjugramam, Tamil Nadu, India

Abstract. Presently, the urban world are switching to rooftop farming with technology support to cope with the increasing food demand and effective utilization of various resources. But the monitoring of farming in high raised building (rooftop) is little challenging. This paper presents an IoT based smart rooftop irrigation system to efficiently manage water dispersal and provides improved urban based rooftop farming productivity. The proposed model regularly monitors soil temperature and moisture level, i.e., rooftop farming are irrigated automatically without physical presence of planter and it also uses smart mobiles for irrigation control. The proposed automatic irrigation farming modal uses sensor technology, communication technology and embedded hardware technology. That is, the atmospheric weather data such as moisture level and water level are collected periodically from different part of rooftop farm area and it is analyzed, which will trigger the switching of water motor/overhead tank pump and send status updates to the planter.

Keywords: Rooftop farming · Urban agriculture · IoT in agriculture · IoT · Smart agriculture

1 Introduction

According to the Commission on Sustainable Agriculture and Climate Change, the world's food demand and its production is progressively observed more as global population grows [1–5]. There is a very strong connection between the growth of living standards, income and wealth in a nation and the demand for food that represents this higher standard of living. This shift in preferences is associated with an increased demand for technology aided land intensive food production sources. The current situation in agriculture is privation of farming land scarcity, and traditional agricultural methods have drawbacks in essential management issues like water resources planning and monitoring, ineffective usage of man power, electrical power etc. [6, 7].

One among the strength of developing countries like India is stable agriculture productivity. It is also a backbone of most of the countries. The feed requirements of the nation are satisfied up to certain level by own agriculture fields. But, the present situation

© IFIP International Federation for Information Processing 2021
Published by Springer Nature Switzerland AG 2021
V. Krishnamurthy et al. (Eds.): ICCIDS 2021, IFIP AICT 611, pp. 52–63, 2021.
https://doi.org/10.1007/978-3-030-92600-7_5

is hasty urbanization of the country i.e. making all agriculture land surface to buildings and reprehensible management of water resources force to import agriculture products from other countries like China, Sri Lanka, Australia, and other Asian & European countries due to urbanization in India and lack of proper productivity management [1, 3, 5]. Due to urbanization the total agriculture land loss is rapid in India. Most of the agriculture lands are now made to concrete high raised buildings and due to which there exist scarcity of land area for irrigation. The statistical graph discussed in counterview portal for agriculture land loss in India is illustrated in Fig. 1. This graph shows the effect of urbanization and in turn confirms the pitfall of agriculture land.

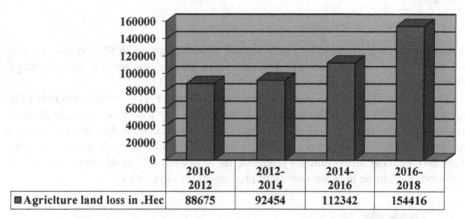

	2010-2012	2012-2014	2014-2016	2016-2018
■ Agriclture land loss in .Hec	88675	92454	112342	154416

Fig. 1. Statistical graph for agriculture land loss in India (between 2010–2018).

Another fundamental concern in traditional agriculture is water resource deployment for farm area and its related maintenance are done manually with less technological aids. As depicted earlier the urbanization also escort to ineffectual usage of water resources which also intern impact on sustainable agriculture. An effectual agriculture system demands a good irrigation surface and effective water resource stream [2, 4, 6–11]. The Table 1 shows the extent of water resources used in India as per survey for the year 2019–2020. But the effective use of these resource is possible increased by technological aid like IoT, Data analytics, Cloud computing, Machine learning etc. [1, 7, 12–15].

The IoT procedure are habitually called as "smart" devices because they have sensors and intricate data scrutiny programs. IoT devices accumulate data via sensors and provide services to the users based on scrutinizes of the data and rendering to user-defined parameters [9, 10, 16]. Refined IoT devices can "learn" by diagnosing patterns in user inclinations and historical user data. An IoT device can become "smarter" as its program regulate to get better its prediction potential so as to augment user practice. The basic foundation of IoT is integration of sensors, actuators, RFID tags, and other communication objects that are connected via Internet to satisfy common goals [11, 13, 17–20]. Internet of Things (IoT) is a link of sensors and connectivity to empower application similar to agriculture ideal irrigation. The preeminent components of IoT are wireless sensor network (WSN). WSN is capable to be the solutions for a huge assortment of applications. The application instances are agriculture monitoring, health monitoring, air

Table 1. Major water sources for irrigation in India.

Water source for irrigation	% share of holdings
Tube wells	39.22%
Canals	25.17%
Wells	19.01%
Other sources	8.34%
Tanks	5.18%
Rain water	3.08%

quality monitoring, weather observing and soil slide monitoring. WSN is syndicates several technologies such as control technology, sensor technology, networking technology, information storage and processing technology [12–14, 21].

The social and economic impact of proposed rooftop based smart agriculture is illustrated in dependency cycle shown in Fig. 2. In addition with food production, implementation of rooftop farming will also provide many non-functional benefits viz. reduces the urban hotness island effect (global warming), reducing greenhouse gases emissions and reducing air pollution, systematic rainwater run-off, freshening and aesthetic surrounding farming area, air quality improvement, biodiversity etc.

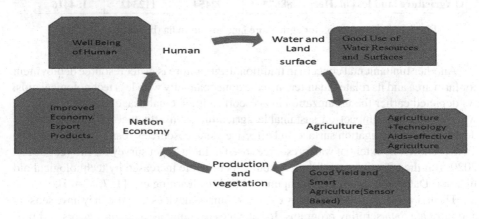

Fig. 2. The dependency cycle of effective rooftop farming system.

The proposed model is very useful to the growth of the nation economy in terms of agriculture production from building rooftops which are generally kept unexploited previously. According to the idea proposed in this paper, the general public can perform vegetation from their building's rooftop or providing it to others in rental basis to earn money.

2 Comparative Analysis

In recent research many researchers have emerge with latest technology aid for efferent agriculture. The major design goals of an IoT system are energy efficiency of nodes, low latency, high throughput, scalability, different topology support, security, safety and privacy of nodes [12, 19, 22]. The data in Table 2 depicts a survey on such study highlights various authors contribution along with key components used in their work, but these research have several pitfalls.

Table 2. Comparative analysis of existing works on smart/precision agriculture.

S. No	Author(s)	Year	Hardware components used	Outcomes
1	Vaishali et al. [1]	2017	Soil moisture sensor, Raspberry-Pi, Blue Term, Motor	Soil moisture level is not intimated to user. It only sends alert SMS to user
2	Sushanth G and Sujatha S [2]	2018	Arduino, Moisture Sensor, Humidity sensor, Temperature sensor, Motion sensor, Relay, GSM module, Motor	User can't able to know whether the water is flown to every corner of the land or not
3	Pushkar Singh and Saikia [3]	2016	Arduino, Temperature sensor, Soil Moisture Sensor, ESP8266 Wifi module, Motor	Website communication is needed to intimating information to user
4	Shweta B et al. [4]	2017	Zigbee, GPRS/3g/4g modules, Wi-Max, Wi-Fi, Bluetooth, soil moisture & temperature sensor	Increases the cost, scalability has to be improved
5	Nageswara Rao and Sridhar B [5]	2018	Raspberry pi 3 – model,LM 358, Temperature sensor, Soil Moisture Sensor, Relay, Buzzer, Motor	It is high in cost and maintenance cost of components is high
6	Kiranmai Pernapati [7]	2018	ESP8266, Moisture Sensor,Humidity sensor, Ultrasonic sensor, Relay, Motor	Sensor information and Water level are not intimated to user
7	Kizito Masaba et al. [10]	2016	ESP8266, Moisture sensor, Temperature sensor, Bluetooth, Motor	It works on Bluetooth technology so it covers only limited region

(continued)

Table 2. (*continued*)

S. No	Author(s)	Year	Hardware components used	Outcomes
8	Hamami Loubna and Nassereddine Bouchaib [22]	2018	Zigbee,EC-5 Soil Moisture Sensor and DS18B20 temperature sensor	Prototype of a smart irrigation system using WSN and ZigBee is developed but accuracy is minimum
9	Le D.N. et al. [20]	2019	Intelligent Data Processing sensors, IoT based SMS, Cloud services	It assesses business model canvas for IoT based startups
10	Ledesma G. et al. [21]	2020	City Maps, Buildings information, IoT Farming Technology	Sustainability Assessment of rooftop farming with respect to technical, economic, environmental, legal and urban factors

Apart from the study performed in Table 2, the present revolutionary irrigation is rooftop farming. But the existing rooftop irrigation system is a manual system which is purely human based, and it requires more labor for its operation. The drawbacks in existing approach are it requires more time, money, labor and it leads to erroneous water management [6, 17]. These drawbacks are overcome by automating the irrigation system using currently available technological advancements.

This proposed work inculcates IoT ideas in smart farming to monitor & automate the tasks involved in the rooftop irrigation system in land and water scarcity vicinity like urban condition. The environmental conditions related to the selected rooftop agricultural area are regularly sensed using different sensors/actuators. The collected data are feed to a cloud server with a decision making system, which directs various components present in the system to work as per the predefined instructions given by the farmer/user.

3 Proposed Work Methodology and Design Model

In the proposed IoT based smart rooftop farming, the farming area is created with the help of sensors/actuators, IoT infrastructure and data analytics capability. The sensors used in for the model includes humidity (DTH11), soil moisture (REES52) sensors. In the system, moisture sensors is submerged into the soil with different locations, which would alert the system about quantity of water content in the soil. Sensors acquire the environmental conditions of the selected area on rooftop and send the collected data to cloud server. If the moisture level is not as much of the amount of water desirable by the plant, the system automate or mechanize the flow of water from an overhead tank unless a threshold value is reached. This make certain that plant has been supplied optimum quantity of water without any physical labor or consumption. The IoT infrastructure provides the environment for the whole system to work, which includes sensors/actuators for data acquisition, wireless technologies like wireless sensor Networks, Wi-Fi/Mobile network etc. for data communication, cloud server for storing and processing of sensed data. The input data are processed according to set of predefined rules and the water level threshold given in Table 3 to handle the working of smart rooftop irrigation system.

In order to make a precise model of the proposed system, the components used in this work are, sensors/actuators (humidity and soil moisture), Raspberry Pi kit, server, smart phone, electrical motor, automatic starter, etc. This proposed work simplifies the risks in irrigation farming system, further it promotes effective water management in rooftop farming area. The detailed architecture diagram is shown in the Fig. 3.

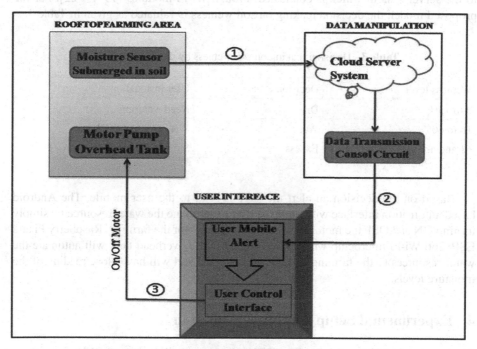

Fig. 3. Schematic diagram of proposed rooftop farming model.

The architecture diagram shown in Fig. 3 explain the working of proposed approach, the entire system is split into three modules viz. rooftop farming area (Sensors/actuators, Overhead tank), Data manipulation (Cloud Server for data handling and processing), User console (Smart phone for regular monitoring and alerts). The soil moisture sensor will sense the moisture range in the soil at regular time interval. The readings from the sensor are regularly monitored and processed to the server. In general, the moisture content in the soil is formulated by the weight of the soil with adequate wetness content in the soil to that of low wetness in the soil.

Calculate the moisture content on a wet-weight basis using the following Eq. 1,

$$\text{Moisture content } (\%) = \frac{W2 - W3}{W2 - W1} * 100 \qquad (1)$$

where,
W1 = initial weight of soil;
W2 = weight of soil and vegetation before drying; and
W3 = weight of soil and vegetation after drying.

The readings from the sensor are handled by the server. The server system makes pronouncement on data and process the data to the user mobile system as an alert. The decision making process is done based on regular monitoring the data from the sensor to the server system which is connected through Wi-Fi modules [15–17, 22]. For the proposed model, the decision making on soil wetness is valuated as given in Table 3.

Table 3. Decision making on soil wetness by the server system.

Wetness level	Decision	User insinuation
0 to 15%	Dry	Need watering
16 to 60%	Wet	Water content is Ok
61 and above	Excess	Stop watering

Based on the decision an alert message is send to the user mobile. The Android based automation interface will guide the user to automate the water resource by simply turning ON and OFF the motor/tank on a single touch for the farming. Raspberry Pi and ESP8266 Wi-Fi microchip which are circuited to the overhead tank will automate the water resources to the farming and the sensor in the soil will have close reading of the moisture levels.

4   Experimental Setup and Implementation

The proposed work used humidity (DTH11) sensor, soil moisture (REES52) sensor, Raspberry Pi and ESP8266 Wi-Fi microchip. The moisture sensor immersed in soil reads moisture level of the soil regularly. This sensor plays major role in this work to notify the level of water content existing in the soil. The core part of processing is performed at server which takes necessary actions based on the pre-defined decision logic. The networking capabilities between different components in the system are established by Raspberry Pi and ESP8266 Wi-Fi microchip kits. The overall functioning of the system is monitored and controlled remotely using android based smart phone. Relay is connected to motor for controlling operation of the motor. The traditional rooftop farming and rooftop farming with technological aid are shown in Fig. 4(a-b).

Fig. 4. The urban rooftop farming. a. Traditional method[1] and b. With technology aid[2]. (Source: [1]www.thehindu.com, [2]economictimes.indiatimes.com)

As shown in Fig. 4, the existing system needs manual monitoring for farming and water sources management. But the proposed approach, employs IoT technology for Rooftop Farming and water management. The working model of proposed IoT based Rooftop Farming along with user notification module are shown in Fig. 5. The Table 4 lists out different functional benefits of using IoT technology in rooftop farming.

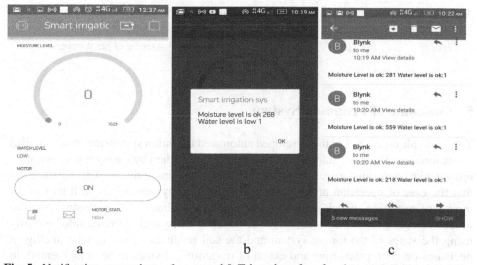

Fig. 5. Notification screenshots of proposed IoT based rooftop farming. a) IoT based mobile monitoring b) Low moisture indication through mobile c) Mailing system for consistent monitoring.

Table 4. Functional benefits of rooftop farming with IoT aid.

Effects category	Functional benefits	
	Effects	Factors which influence effects
Rooftop mitigation impacts and water resource utilization through IoT adaptation impacts	Reducing the Urban Heat Island Effect	– Plant leaf area index – Green surface coverage
	Reducing Greenhouse gases emissions	– Green surface coverage (insulation) – Plant yield (volumes of food produced locally)
	Carbon storage and sequestration	–Soil coverage and depth, plant species used, degree of permanent cover and harvest
	If it rains then reducing rainwater run-off	– Soil depth –Plant leaf area index – Green surface coverage
	Biodiversity improvement	– Input of organic matter – Depth and composition of the soil –Variety of plant species and degree of permanent cover
	Reducing air pollution	– Plant leaf area index –Degree of permanent cover
Developmental benefits	Reducing food insecurity	–Intensity of growing – Plant yield
	Improved living environment	–Degree of plant cover

5   Outcome of Proposed System

The principle objectives of the developed automated irrigation system are to accomplish operational intelligence, automation and independence thereby easing the stress often related with conformist irrigation performs. The innovations of this work lies in the point that the ease of operation of the developed automated system comes with its various modes of operation, which gives the user convenient choices of operation [18]. The modes comprise the full automation mode whereby the system operates autonomously using the states of the timing system and the soil feedback sensor to take intelligent decisions on appropriate time and extent of irrigation activities to be implemented. In addition, a user could send a command remotely via mobile phone to the field system to start or stop irrigation. Furthermore, users have the privilege of querying the system remotely to acquire response on the soil tangible time settings. Upon receiving such command via the Short Message Service (SMS), the system queries the soil feedback sensor and sends the soil moisture readings to the authorized user(s). The system can also be made to operate in the manual mode relying on human control during certain periods such as the start up or shut down based on the environment and nature of irrigation activities to be carried out. For the purpose of proposed work, only two moisture sensors are been used in the experiments. The features of the soil moisture sensor include: current 3.5 mA, input voltage power supply (3.3 V–3.5 V), output voltage power supply (0–4.2 V), size (60 × 20 × 5 mm).

The soil moisture sensor comprises of two probes which conduct current into the soil and thereafter reads the electrical resistance so as to regulate the moisture level. Extraordinary moisture content in the soil often triggers high conductivity of current by the soil and vice-versa. The output voltage from the moisture sensor which is immersed in soil is then amplified and sent to the micro controller where it is transformed to digital format using Analog to Digital Converter (ADC). On the controller, the voltage is linked to the threshold value pre-set on it, if the voltage measured is less than the threshold value, the micro controller activates the relay which turns ON the pump for irrigation activities. On the other hand, if the measured value exceeds or equal to the threshold, it implies there is no need for irrigation hence system remains in the idle state. Since the system is automated, manpower is also reduced. Such type of farming (IoT based rooftop farming) have dependency cycle for every human, where such farming will not only improve the vegetation system in urban sector but, it also improves the single-to -nation's revenue.

Table 5. Experimental setup and specifications of proposed model.

Experimental setup		
Month and Year	April 2018 to December 2018	Duration - Nine months
Farming model	Roof Top	With Green Agro Net
Farming area	3600 sq.ft	60 × 60 sq.ft
Farming kit	Organic Potting Soil Mix Fertilizer, Red and regional soil mix, Grow Bags	Grow bags size:16 × 16 × 30 cm
Vegetation	Tomato, green chili, better gourd and ladyfinger	Seasonal Vegetation's
Climate	Summer –Rainy - Winter	Summer –4 monthsRainy-3 months, Winter-2 months
Sensors used	1. Raspberry pi 4 2. Soil Moisture Meter Soil Humidity Sensor Water Sensor UNO 3. NodeMCU ESP8266 CH340 WIFI Node	Model Number: BE-000030Model B CH340 WIFI Starter Kit
Output of yield	Vegetation by seasonal conditions (Near approx. in kilo grams)	8 kg. of tomato 5 kg.of green chilli 6 kg (3 + 3) of bitter gourd and ladyfinger

The details about experimental setup and specifications of the components used in the proposed work are given in Table 5.

Experimental tryout were executed from April 2018 to December 2019 on the rooftop of a public lodging building in the city of Villupuram (Tamil Nadu, India). Tamil Nadu is a representative state were year-round open-air rooftop farming practices can be accomplished due to promising climatic conditions. The trial crops were grownup in a communal garden executed on the 334-m^2 terrace of the 3rd floor of the building. This current rooftop farming and their latent to be low-cost choices for self-managed rooftop gardens. Soil production was also prepared on wooden ampules where plants were grownup on commercial soil with manure and fertilizers. IoT based tank water was used for irrigation in all the systems meanwhile design does not consider any rainwater harvesting system. The crops like tomato, green chili, better gourd and ladyfinger were selected for vegetation since they are all seasonal and throughout year available crops.

6 Conclusion

The research work presented in this paper provides IoT based smart agriculture solution for irrigation of rooftop vegetation in different climates. This automatic irrigation system uses sensor technology with micro controller to create smart switching mechanism and this model displays simple switching tool of water motor/pump using sensor from any part of ground or soil by sensing moisture existing in the soil. This work used humidity, soil moisture sensors, Raspberry Pi and ESP8266 Wi-Fi microchip to make smart switching of various components to ensure automatic water dispersal in rooftop farming area. The main advantages of proposed work are optimum use of water resources, minimum use of human labor, energy saving, cost saving, automation etc. than traditional irrigation techniques. The applications of rooftop farming will up-rise the creative green roofs conglomerate food production with natural sustainability, such as abridged rainwater run-off, temperature aids such as latent reduction of heating and cooling necessities, biodiversity, enhanced aesthetic significance and also air quality in urban vicinity.

References

1. Vaishali, S., Suraj, S., Vignesh, G., Dhivya, S, Udhayakumar, S.: Mobile integrated smart irrigation management and monitoring system using IoT. In: International Conference on Communication and Signal Processing (ICCSP), pp. 2164–2167 (2017)
2. Sushanth, G., Sujatha, S.: IoT based smart agriculture. In: International Conference on Wireless Communications, Signal Processing and Networking (WiSPNET), pp. 1–4 (2018)
3. Pushkar S., Sanghamitra, S.: Arduino-based smart irrigation using water flow sensor, soil moisture sensor, temperature sensor and ESP8266 WiFi module. IEEE Region 10 Humanitarian Technology Conference (R10-HTC), pp. 1–4 (2016)
4. Shweta, B.S., Dhanashri, H.: IoT based smart irrigation monitoring and controlling system. IEEE International Conference on Recent Trends in Electronics, Information & Communication Technology (RTEICT), pp. 815–819 (2017)
5. Nageswara, R.R., Sridhar, B.: IoT based smart crop-field monitoring and automation irrigation system, 2018. In: 2nd International Conference on Inventive Systems and Control (ICISC), pp. 478–483 (2018)

6. Prathibha, S.R., Anupamab, H., Jyothi, M.P.: IoT based monitoring system in smart agriculture. In: International Conference on Wireless Communications, Signal Processing and Networking (WiSPN), pp. 81–84 (2017)
7. Kiranmai, P.: IoT based low cost smart irrigation system. In: International Conference on Inventive Communication and Computational Technologies (ICICCT), pp. 1312–1315 (2018)
8. Priyanka, P., Sonal, M., Kartikee, D., Sushmita, M., Deepali, J.: Smart water dripping system for agriculture/farming. In: International Conference for Convergence in Technology (I2CT), pp. 659–662 (2017)
9. Arjun, P., Stephenraj, S., Naveen Kumar, N., Naveen Kumar, K.: A Study on IoT based smart street light systems. In: IEEE International Conference on System, Computation, Automation and Networking (ICSCAN), pp. 1–7 (2019)
10. Kizito, M., Amini, N., Tahaselim, U.: Design and implementation of a smart irrigation system for improved water-energy efficiency. In: 4th IET Clean Energy and Technology Conference (CEAT), pp. 1–5 (2016)
11. Odara, S., Khanand, Z., Ustun T.S.: Integration of precision agriculture and smart grid technologies for sustainable development. In: IEEE International Conference Technological Innovation in ICT for Agriculture and Rural Development (TIAR), pp. 84–89 (2015)
12. Xu, L.D., He, W., Li, S.: Internet of Things in Industries: a survey. IEEE Trans. Ind. Inf. **10**(4), 2233–2243 (2014)
13. Sales, N., Remédios, O., Arsenio, A.: Wireless sensor actuator system for smart irrigation on the cloud. In: IEEE 2nd World Forum on Internet of Things (WF-IoT), pp. 693–698 (2015)
14. Ayaz, M., Ammad-Uddin, M., Sharif, Z., Mansour, A., Aggoune, E.-H.M.: Internet-of-Things (IoT)-based smart agriculture: toward making the fields talk. IEEE Access **7**, 129551–129583 (2019)
15. Elavarasan, D., Vincent, P.M.D.: Crop yield prediction using deep reinforcement learning model for sustainable agrarian applications. IEEE Access **8**, 86886–86901 (2020)
16. Paucar, L.G., Diaz, A.R., Viani, F., Robol, F., Polo, A., Massa, A.: Decision making for smart irrigation by means of wireless distributed sensor. In: IEEE 15th Mediterranean Microwave Symposium (MMS), pp. 1–4 (2015)
17. Zaier, R., Zekri, S., Jayasuriya, H., Teirab, A., Hamaza N., Al-busaidi, H.: Design and implementation of smart irrigation system for groundwater use at farm scale. In: 7th International Conference on Modelling, Identification and Control (ICMIC), pp. 1–6 (2015)
18. Ghosh, M.: Climate-smart agriculture, productivity and food security in India. J. Dev. Policy Pract. **4**(2), 166–187 (2019)
19. Elijah, O., Rahman, T.A., Orikumhi, I., Leow, C.Y., Hindia, M.N.: An overview of Internet of Things (IoT) and data analytics in agriculture: benefits and challenges. IEEE Internet Things J. **5**(5), 3758–3773 (2018)
20. Le, D.N., Tuan, L.L., Tuan, M.N.D.: Smart-building management system: an Internet-of-Things (IoT) application business model in Vietnam. Technol. Forecast. Soc. Chang. **141**, 22–35 (2019)
21. Ledesma, G., Nikolic, J., Pons-Valladares, O.: Bottom-up model for the sustainability assessment of rooftop-farming technologies potential in schools in Quito, Ecuador. J. Clean. Prod. **274**, 122993 (2020)
22. Hamami, L., Nassereddine, B.: Towards a smart irrigation system based on Wireless Sensor Networks (WSNs). In: International Conference of Computer Science and Renewable Energies (ICCSRE), pp. 433–442 (2018)

Enhanced Ant Colony Optimization Algorithm for Optimizing Load Balancing in Cloud Computing Platform

A. Daniel$^{(\boxtimes)}$, N. Partheeban, and Srinivasan Sriramulu

Galgotias University, Greater Noida, UP, India
{n.partheeban,s.srinivasan}@galgotiasuniversity.edu.in

Abstract. Load balancing among virtual machines (VMs) is significant for conveying the cloud services in advanced manner with least cost paid and all-out time procured to convey the services. In this research work, the different research holes for load balancing optimization in the past writing have been exhibited, which should be tended to for taking care of the load balancing issue in cloud environment. In present work, Hybrid methodology-based asset provisioning and load balancing framework for work processes execution has been proposed to enhance the usage of VMs with uniform load conveyance. The proposed framework depends on binary ant colony optimization algorithm and the results proved the superiority of them over the compared methods.

Keywords: Cloud · Load balancing · Ant colony · Virtual machine

Load balancing is important for enhanced utilization of cloud resources (processors, memory, circles) and to get superior of the machines. The resources are allotted and used by VMs facilitating on physical machines. At the point when undertakings are booked on VMs there may emerge a circumstance that a portion of the VMs are over utilized, while others stay under-used.

Load balancing techniques are utilized to ensure that each machine in the cloud datacenter performs around the equivalent number of assignments anytime of time. In cloud computing, the requests of client are exceedingly unique in nature and multi-tenure requires confining diverse clients from one another and from the cloud foundation. Distinctive heuristic and metaheuristic techniques have been utilized for circulating load among accessible VMs and to get ideal execution of VMs.

The computational expense of metaheuristics is higher than heuristics techniques since they have bigger inquiry space and their hunt criteria depends on guided arbitrary look to get scan answer for the given planning issue. Heuristics are utilized to diminish the scan space for metaheuristics to improve their combination rate to achieve the ideal arrangement in least conceivable time. Notwithstanding, the issue of load balancing is multi-objective, with the end goal that the essential target is to convey occupations or errands among accessible resources so the relative lop- sidedness is limited; and the optional goal is to boost the use of resources while lessening makespan and cost.

© IFIP International Federation for Information Processing 2021
Published by Springer Nature Switzerland AG 2021
V. Krishnamurthy et al. (Eds.): ICCIDS 2021, IFIP AICT 611, pp. 64–70, 2021.
https://doi.org/10.1007/978-3-030-92600-7_6

In this examination paper, a framework for resource provisioning and load balancing has been proposed and actualized. The proposed framework depends on binary ant colony optimization algorithm to accomplish its ideal execution as far as makespan and cost. A framework for Workflow's execution has been proposed to give ideal load to underutilized VMs and advance their use.

1 Literature Review

The load balancing techniques for cloud computing with various heuristic and meta-heuristic techniques explicitly for load balancing optimization have been looked into. Different load balancing techniques have been recommended by the analysts. The exploration holes have been found amid the past writing study. [1] proposed Load Balancing Ant Colony Optimization (LBACO) algorithm for balancing the load crosswise over whole framework and minimize makespan.

The outcomes are observed to be superior to fundamental ACO and FCFS algorithms. The undertakings have been viewed as commonly autonomous with no priority require-ment, nonprimitive and calculation serious. The creators considered free undertakings as contribution with no priority and spending requirements among them. Additionally, ACO parameters have been instated arbitrarily.

[2] proposed load balance improved min-min scheduling (LBIMM) algorithm which expanded asset usage and diminished makespan. Be that as it may, due dates regarding execution time and cost have not been given significance in their work. Further, [3] thought about makespan of ACO-based cloud task scheduling with FCFS and Round Robin algorithms and presumed that ACO beat the two partner algorithms in any case, reliance limitations among errands have not been considered, [4] proposed ACO-based load balancing optimization to allot balanced load of errands among VMs.

They have taken just finish time algorithm examination with predetermined number of VMs (up to 4) and no other measurement has been utilized to assess its execution. [5] proposed and actualized a dynamic load the executive's algorithm for viable circu-lation of solicitations among VMs. Be that as it may, the reaction time expanded when underloaded VMs were checked.

[6] proposed multi-objective ACO (MO-ACO) based undertaking scheduling and load balancing algorithm with makespan and cost optimization. By and by, non-pre-emptive scheduling has been done on autonomous undertakings which are not appro-priate for cloud environment. Since, load balancing alludes to managing hubs or VMs which are used beneath their minimum limit. This outcomes in wastage of computing asset if not used to their ideal point of confinement.

At the point when VMs are over utilized, the all-out time taken by a VM to finish every one of the assignments apportioned to it, i.e., its makespan, additionally increments. Be that as it may, when the VMs are underutilized, in spite of the fact that the makespan decline however it results in expanded expense of asset usage as the accessible resources (VMs) are not broadly used bringing about their wastage. Subsequently, there is a need to balance the load over the VMs with the goal that both the makespan and cost parameters must be controlled and balanced. The abatement in makespan must not result in increment of expense of asset use and the other way around.

In past research works, load balancing optimizations has been finished utilizing diverse metaheuristics yet in practically all the proposed techniques, the meta-heuristics are furnished with irregular seed for their parameters [1]. The arbitrary introduction of metaheuristic parameters can corrupt the execution of the method.

The greater part of the load balancing approaches proposed in writing spotlight on minimizing just the execution time yet don't consider the execution cost [7]. Cost is a significant parameter as the cloud computing services are conveyed on pay per use dependent on web. What's more, the priority request of assignment execution with requirement on execution time and cost has not given due significance in past works to such an extent that the errands which don't satisfy these limitations (due dates) ought not to be executed [8–10].

2 Proposed Method

2.1 Solution Construction

In BAS, artificial ants construct solutions by walking on the mapping graph as described in Fig. 1 At every iteration i, a number of ants n_a cooperate together to search in the binary solution domain, each ant constructs its solution by walking sequentially from node 1 to node $n+1$ on the routing graph. At each node, ant either selects the upper path $i0$ or the selects lower path i_1 to walk to the next node $i+1$. $i0$ Selecting means $x_i = 0$; and i_1 selecting means $x_i = 1$. The selecting probability is dependent on the pheromone distributed on the paths.

$$p_{is}(t) = \tau_{is}(t), i = 1, ..., n, s \in \{0, 1\} \tag{1}$$

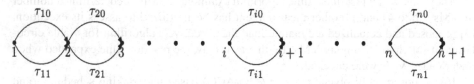

Fig. 1. Routing diagram for ants in BAS

The solutions constructed by the ants may not be feasible when tackling the constrained binary optimization problems. A solution repair operator is incorporated to transfer the infeasible solutions to the feasible domain.

2.2 Pheromone Update

Initially, BAS set all the pheromone values as $\tau_{is}(0) = 0.5$, which is the same as that of HCF [8], but uses a simplified pheromone update rule:

$$\tau_{is}(t+1) \leftarrow (1 - \rho)\tau_{is}(t) + \rho \sum_{x \in S_{upd} \mid is \in x} w_x \tag{2}$$

here S_{upd} is the set of solutions to be intensified w_x are explicit weights for each solution $x \in S_{upd}$

$$0 \leq w_x \leq 1 \quad \rho = \rho 0$$

which satisfying and
$\sum_{\in S_{upd}} w_x = 1$ is the evaporation parameter, which is set initially as, but decreases as $\rho \leftarrow 0.9\rho$ every time the pheromone re-initialization is performed.

S_{upd} consists of three components, they are: the global best solution S^{gb}, the iteration best solution S^{ib}, and the restart best solution S^{rb}.

Different combinations of w_x are implemented according to the convergence status of the algorithm. The convergence status is monitored by a convergence factor cf, which is defined as

$$cf = \frac{\sum_{i=1}^{n} | \tau_{i0} - \tau_{i1}|}{n} \tag{3}$$

Under this definition, when the algorithm is initialized with $\tau_{is}(0) = 0.5$ for all the paths, $cf = 0$, while when the algorithm gets into convergence or premature, $|\tau_{i0} - \tau_{i1}| \to 1$, thus that $cf \to 1$.

Table 1. Pheromone update strategy for BAS

	$cf < cf_1$	$cf \in (cf_1, cf_2)$	$cf \in (cf_2, cf_3)$	$cf \in (cf_3, cf_4)$	$cf \in (cf_4, cf_5)$
w_{ib}	1	2/3	1/3	0	0
w_{rb}	0	1/3	2/3	1	0
w_{gb}	0	0	0	0	1

Table 1 describes the pheromone update strategy in different value of cf, where $w_{ib} w_{rb}$ *and* w_{gb} are the weight parameters for $S^{ib} S^{rb}$ *and* S^{gb} respectively, $cf, i = 1, \cdots, 5$ are threshold parameters within the range of [0,1].

In BAS, once $cf > cf5$, the pheromone re-initialization procedure is performed according to S^{gb}:

$$\begin{cases} \tau_{is} = \tau H & if \ is \in S^{gb} \\ \tau = \tau L & otherwise \end{cases} \tag{4}$$

these two parameters satisfying $0 < \tau L < \tau H < 1$ *and* $\tau L + \tau H = 1$. This kind of pheromone re-initialization will focus more on the previous search experience rather than a total redo of the algorithm.

3 Results and Discussion

To balance load among the VM's in cloud service is a composite task that needs to be done. The existing and proposed method performance is measured by means of increasing the number of VMs in cloud. Figure 1 shows the results of the presented method in terms of average make span.

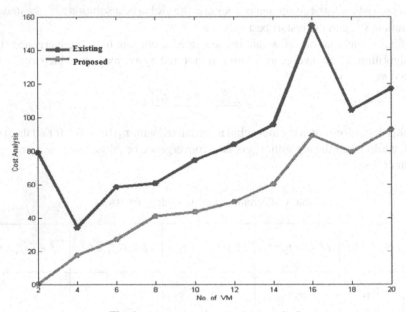

Fig. 2. Average makespan (ms) analysis

Figure 2 shows comparison of existing and proposed method average makespan analysis while running the Workflow of genome in the framework of presented method. For Ligo workflow execution, while 2 VMs are employed as resource for the existing method, the ms achieved the cost analysis of 80 and decreased when deployed by 4 VMs and subsequently increased till 14 number of VMs, finally it increases drastically near 160 and end near 120 when the number of VMs are 20.

When 2 VMs are employed as resource for the existing method, the ms achieved the cost analysis is nearby zero and increased shortly when deployed by 4 VMs and subsequently decreased in 14 number of VMs, finally it increased a little bit near 100 when the number of VMs are 20. From the makespan results computed during execution of three scientific workflows on the proposed are better than existing approach.

Figure 3 shows comparison of existing and proposed method average time analysis while running the Workflow of Cybershakein the framework of HDD-PLB. When 2 VMs are employed as resource for the existing method achieved the time analysis of 2.5 and decreased shortly when deployed by 4 VMs and subsequently decreased in 14 number of VMs, finally it increased a little bit near 1.5 when the number of VMs are 20. While 2 VMs are employed as resource for the proposed method, the time analysis of

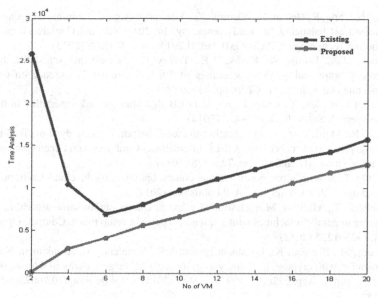

Fig. 3. Average cost (Rs) analysis

80 and increases when deployed by 4 VMs and subsequently increased till 14 number of VMs, finally it increases drastically near 1 when the number of VMs are 20. From the time analysis results computed during execution of three scientific workflows on the proposed are better than existing approach. The results of proposed method show consistent pattern of low cost incurred with small number of VMs and then rises slowly with increased VMs.

4 Conclusion

Load balancing techniques are utilized to ensure that each machine in the cloud data centre performs around the equivalent number of assignments anytime of time. In cloud computing, the requests of client are exceedingly unique in nature and multi-tenure requires confining diverse clients from one another and from the cloud foundation. The resources are allotted and used by VMs facilitating on physical machine. The proposed framework depends on binary ant colony optimization algorithm. From the time analysis results computed during execution of three scientific workflows on the proposed are better than existing approach. The results of proposed method show consistent pattern of low cost incurred with small number of VMs and then rises slowly with increased VMs.

References

1. Li, K., Xu, G., Zhao, G., Dong, Y., Wang, D.: Cloud Task Scheduling based on load balancing ant colony optimization. In: IEEE 2011 Sixth Annual Chinagrid Conference (ChinaGrid), pp. 3–9 (2011)

2. Chen, H., Wang, F., Helian, N., Akanmu, G.: User-priority guided min-min scheduling algorithm for load balancing in cloud computing. In: 2013 National Conference on Parallel Computing Technologies, PARCOMPTECH 2013, pp. 1–8. IEEE (2013)
3. Tawfeek, M.A., El-Sisi, A., Keshk, A.E., Torkey, F.A.: Cloud task scheduling based on ant colony optimization. In: Proceedings of 8th International Conference on Computer Engineering and Systems (ICCES), pp. 64–69 (2013)
4. Xue, S., Li, M., Xu, X., Chen, J.: An ACO-LB algorithm for task scheduling in the cloud environment. J. Softw. **9**(2), 466–473 (2014)
5. Panwar, R., Mallick, B.: Load balancing in cloud computing using dynamic load management algorithm. In: Proceedings of IEEE International Conference on Green Computing and Internet of Things (ICGCIoT), pp. 773–778 (2015)
6. Guo, Q.: Task scheduling based on ant colony optimization in cloud environment. In: Proceedings of AIP Conference, AIP Publishing (2017)
7. Keskinturk, T., Yildirim, M.B., Barut, M.: An ant colony optimization algorithm for load balancing in parallel machines with sequence-dependent setup times. Comput. Operat. Res. **39**(6), 1225–1235 (2012)
8. Elhoseny, M., Shankar, K., Lakshmanaprabu, S.K., Maseleno, A., Arunkumar, N.: Hybrid optimization with cryptography encryption for medical image security in internet of things. Neural Comput. Appl. **32**(15), 10979–10993 (2018). https://doi.org/10.1007/s00521-018-3801-x
9. Karthikeyan, K., et al.: Energy consumption analysis of Virtual Machine migration in cloud using hybrid swarm optimization (ABC–BA). J. Supercomput. **76**(5), 3374–3390 (2018). https://doi.org/10.1007/s11227-018-2583-3
10. Shankar, K., Elhoseny, M., Chelvi, E.D., Lakshmanaprabu, S.K., Wu, W.: An efficient optimal key based chaos function for medical image security. IEEE Access **6**, 77145–77154 (2018)

Captioning of Image Conceptually Using BI-LSTM Technique

Thaseena Sulthana, Kanimozhi Soundararajan, T. Mala, K. Narmatha[(✉)], and G. Meena

College of Engineering, Anna University, Chennai, India

Abstract. Due to the fact of increase in amount of video data each day, the need for auto generation of captioning them clearly is inevitable. Video captioning makes the video more accessible in numerous ways. It allows the deaf and hard of hearing individuals to watch videos, helps people to focus on and remember the information more easily, and lets people watch it in sound- sensitive environments. Video captioning refers to the task of generating a natural language sentence that explains the content of the input video clips. The events are temporally localized in the video with independent start and end times. At the same time, some events that might also occur concurrently and overlap in time. Classifying the events into present, past and future as well as separating them based on their start and end times will help in identifying the order of events. Hence the proposed work develops a captioning system that clearly explains each visual feature that is present in the image conceptually. The Blended-LSTM (Bl- LSTM) model with the help of Xception based Convolution Neural Network (CNN) with Fusion Visual Captioning (FVC) system achieves it with the BLEU score of 75.9%.

Keywords: Blended-LSTM · Fusion visual captioning · CNN · Sports video

1 Introduction

Video captioning is multimedia analysis which is used to generate a natural language sentences by understanding the given video by considering the events.

Taking place in the video. It creates a great impact in computer vision. Automatic video caption generation [9] includes the understanding of many background concepts. It also includes the detection of every occurrence in the video such as objects, actions taking place in the video, scenes taking places, person to person relations in the context of the video, person to object relations in the video and the temporal order of the events. Video captioning also requires translation [7] of the extracted visual information and grammatically correct natural language description.

Conceptual translation plays major role in captioning where most of the misconception occurs. To overcome that pipeline based methods comes into existence as in [12], where visual features are extracted using machine learning algorithm. Extracted features are indexed with words in the vocabulary for framing as sentences using some

© IFIP International Federation for Information Processing 2021
Published by Springer Nature Switzerland AG 2021
V. Krishnamurthy et al. (Eds.): ICCIDS 2021, IFIP AICT 611, pp. 71–77, 2021.
https://doi.org/10.1007/978-3-030-92600-7_7

neural network techniques [6]. So framing of words has to be consummate conceptually, which happens only when there is extraction of required features. Feature extraction is effective when there is suitable region [1] inside the image is selected. Hence Region of Interest (ROI) is determined based on edge detection, contour, histogram, etc.

Conceptual captioning can be done only when there is expected regions features are extracted in a frame. Thus the focus of the proposed work is to select keyframes meaningful which leads to proper feature extraction as well as frame sentence conceptually.

2 Related Works

Recently the auto captioning of images based on the visual information available within it has become major research focus. For producing captions, screening of features is to be done from the initial stage (Keyframe generation) itself. Visual features are more in videos which may be missed if we consider the entire part. So there is a need to convert video into frames [10] for further processing. Selecting a frame from the entire set of frames that represent more visual features is again a huge task. One method for detecting the Keyframe is based on adaptive threshold [4] method. In this threshold value is calculated for a specific region, so that threshold values vary for different regions in a single image. Another method [2] based on HSV histogram, which transfer high-dimensional abstract video image into quantifiable low dimensional data. It results with more appropriate keyframes with low redundancy rate.

Extracting features from the selected keyframes is again difficult task as there is a chance of selecting inappropriate regions. Various methods have been proposed for selecting the regions such as iteration algorithm, region growing and edge detection [6] and so on. In many of these works manual declaration of initial seed is needed which again may create low accuracy. To overcome these issue neural based models like encoder-decoder [5], multimodal layered, etc. has been proposed. In which major focus is towards location as well as fixed size region, where there is no need for the manual interruption. The main contribution of this work is as follows:

(i)   Generating Keyframe from the video using Shot based Adaptive Threshold technique (SAT),
(ii) Extracting selective features from the selected Keyframe using Xception based on CNN,
(iii) Caption is produced with the extracted features using Fusion Visual Captioning (FVC) model.

3 Fusion Captioning Model

The Fusion Visual Captioning (FVC) model consists of important features of CNN and RNN together. Initially the videos are converted into frames for the clear extraction of object features. The frames are generated ate the rate of 20 fps for the video of length 3 min. The keyframe is selected using proposed Shot based Adaptive Threshold (SAT) method as shown in Table 1 (Fig.1).

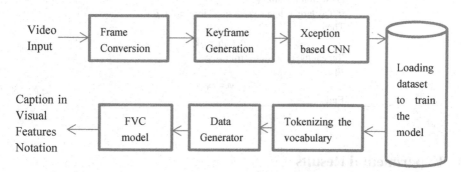

Fig. 1. System design for proposed Bl-LSTM caption model.

Next is the visual feature extraction module in which Xception based CNN is employed. CNN is best known visual feature extractor model [14] that fetches the vectors values based on convolution method [15]. Flickr8k_text dataset is used for captioning which consist of five possible labels for each image. In order to integrate the pre-trained Xception based CNN with Flickr dataset last layer is removed to get 2048 feature vector.

In SAT algorithm (Table 1) threshold $\varnothing_t$ is calculated with mean of the neighborhood pixels subtracted with the constant $\Delta\varnothing$, whereas constant $\Delta\varnothing$ is calculated by subtracting the weighted sum of the neighborhood pixels. The result of the SAT algorithm is the set of keyframes along with major the features incorporated for further processing.

During training phase, a dictionary is created that maps the images used in the dataset along with its descriptions. A unique index value using tokenizer function [11] is assigned to make matching process easier. Display pattern is created with a starting and ending identifier to each word as (start, object, preposition, action, sport and end). The vector value from the last fully connected layer of Xception based CNN is given to the input layer of the LSTM. To make the training process easier for over 10,000 images the data generator module is used. It creates batches using the input feature vector with its text and corresponding predicted output sequence. Finally, the decoder in FVC model embeds the result from previous module with the LSTM for building the expected captions for each image.

Table 1. Proposed SAT algorithm.

Proposed SAT Algorithm
Variables: F_{in} and F_e are the initial and final pixel values of selected region. T is the average pixel value.
Convert pixel values in region P to binary region using the threshold $\emptyset_t$
Assume N is the sum of the number of the non- zero pixels within F_{in} and F_e
If T is larger than a predefined threshold, β
Then
$$\emptyset_t = \emptyset_t + \Delta\emptyset$$
Repeat the procedure from step 1
Else
$$\emptyset = \emptyset_t$$
End

4   Experimental Results

The input to the Bl-LSTM model is given in the form of video due to the fact that future work is focused towards captioning of video sequences. Thus the keyframe is selected in the sense that shows maximum visual representation of entire frame using algorithm in Table 1. Highly featured frame is given as input for the Xception based CNN module which extract 2048 vector values for each image with a dense layer. The dimension is reduced to 256 nodes as in [3] which help for embedding the words towards LSTM layer. Totally of about 7822 words are there in the developed vocabulary. The maximum length of description is fixed to 32 as there are five possible captions for each image. The last layer in the FVC model have nodes (256) equal with that of size (7822) of the generated vocabulary.

Initially the proposed model predicts the caption as per the visual information presents in the frame as shown in Fig. 2. Effectiveness of the system is evaluated based on the BLEU score as in Eqs. (1), (2) [13] that compare the true captions with the predicted caption and generate the final score.

$$BLEU = \min\left(1, \frac{final_{length}}{Ref_{length}}\right)\left(\Pi_{i=1}^{n}precision_i\right)^{\frac{i}{n}} \tag{1}$$

$$Precision_i = \frac{\Sigma_{snt\in cand-corpus}\Sigma_{i\in snt}\min\left(m_{cand}^i, m_{ref}^i\right)}{W_t^i = \Sigma_{snt*\in cand-corpus}\Sigma_{i*\ \in\ snt*,m_{cand}^{i*}}} \tag{2}$$

Where n is the size, m_{cand}^i the count of i-gram in candidate, m_{ref}^i is the count of i-gram of reference, w_t^i is the total i-grams in candidate conversion.

In Fig. 3 x-axis represents 21 possible captions generated using the proposed Bl-LSTM model and y-axis is the corresponding BLEU score of good vs bad captions of the tested sample images. Good as well as bad captions are differentiated in the graph using two

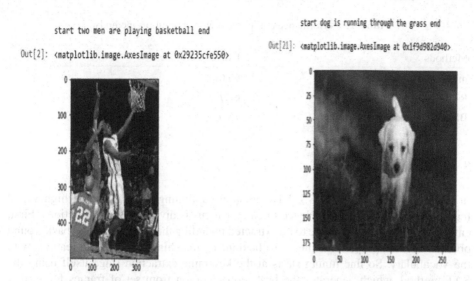

Fig. 2. Examples of final output generated by BL-LSTM caption model

indicators. First with BLEU score, that produces as resultant value of Eq. (1). Second, using true vs predicted caption developed by the proposed model. All the generated captions scores are lies between 7.0–8.0 BLEU values, which resemble the improvement in the Bl-LSTM system performance.

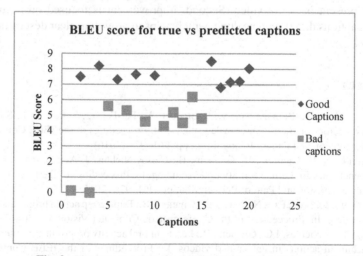

Fig. 3. Final score values for predicted good vs bad captions.

The comparative performance result of our Blended-LSTM (Bl-LSTM) model with other state-of-art methods in Flicker30k dataset is given in Table 2. The BLEU value shows the efficiency of proposed Bl-LSTM model in predicting good (true matching with predicted) captions.

Table 2. Performance comparison with state-of-art methods.

Methods	BLEU score
m-RNN	60.0
R-LSTM	67.7
Bl-LSTM	75.9

5 Conclusion

In this paper, the proposed Bl-LSTM aims at captioning the frame with high visual features representation in it. Hence the major focus is divided into two sections. First, entire features of the frame are to be extracted including the foreground and background objects. Second, proper indexing must be done by matching the extracted features with the vocabulary. So fine tuning starts at the keyframe extraction stage itself using the SAT method, which generates the best representation from set of frames for further processing. Xception based CNN model helps in fetching the highlighted information from the key frames. Next the FVC which is a fusion model that combines meaningfully matched captions among the five possible captions to each keyframe. Efficacy of the proposed model is evaluated using the BLEU score and comparative study is done as well with state-of- art methods. Final BLEU score shows that, proposed Bl-LSTM model outperforms other methods by resulting with 75.9% score value.

Future enhancement is focused towards captioning including visual features of all the events present in entire video. Second, improvement is focused on accuracy for generated captions that should be done, which helps in providing clear description about the video.

References

1. Alahi, A., Goel, K., Ramanathan, V., Robicquet, A., Fei-Fei, L., Savarese, S.: Social lstm: human trajectory prediction in crowded spaces. In: Proceedings of the IEEE Conference on Computer Vision and Pattern Recognition, pp. 961–971 (2016)
2. Caba Heilbron, F., Escorcia, V., Ghanem, B., Carlos Niebles, J. Activitynet: a large-scale video benchmark for human activity understanding. In: Proceedings of the IEEE Conference on Computer Vision and Pattern Recognition, pp. 961–970 (2015)
3. Escorcia, V., Heilbron, F.C., Niebles, J.C., Ghanem, B.: Daps: deep action proposals for action understanding. In: Proceedings of the Conference on Computer Vision, pp. 768–784 (2016)
4. Heilbron, F.C., Niebles, J.C., Ghanem, B.: Fast temporal activity proposals for efficient detection of human actions in untrimmed videos. In: Proceedings of the IEEE Conference on Computer Vision and Pattern Recognition, pp. 1914–1923 (2016)
5. Lee, S., Kim, I.: Multimodal feature learning for video captioning. Math. Prob. Eng. **2018**, 1–8 (2018)
6. Pirsiavash, H., Ramanan, D.: Parsing videos of actions with segmental grammars. In: Proceedings of the IEEE Conference on Computer Vision and Pattern Recognition, pp. 612–619 (2014)

7. Rahman, T., Xu, B., Sigal, L.: Watch, listen and tell: multi- modal weakly supervised dense event captioning. In: Proceedings of the IEEE/CVF International Conference on Computer Vision, pp. 8908–8917 (2019)

8. Xu, H., Li, B., Ramanishka, V., Sigal, L., Saenko, K.: Joint event detection and description in continuous video streams. In Proceedings of the IEEE Winter Conference on Applications of Computer Vision (WACV), pp. 396–405 (2019)

9. Yao, T., Mei, T., Rui, Y.: Highlight detection with pairwise deep ranking for first-person video summarization. In: Proceedings of the IEEE Conference on Computer Vision and Pattern Recognition, pp. 982–990 (2016)

10. Yao, T., Pan, Y., Li, Y., Qiu, Z., Mei, T.: Boosting image captioning with attributes. In: Proceedings of the IEEE International Conference on Computer Vision, pp. 4894–4902 (2017)

11. Young, P., Lai, A., Hodosh, M., Hockenmaier, J.: From image descriptions to visual denotations: new similarity metrics for semantic inference over event descriptions. Trans. Assoc. Comput. Linguist. **2**, 67–78 (2014)

12. Yu, H., Wang, J., Huang, Z., Yang, Y., Xu, W.: Video paragraph captioning using hierarchical recurrent neural networks. In: Proceedings of the IEEE Conference on Computer Vision and Pattern Recognition, pp. 4584–4593 (2016)

13. Zhao, S., Ding, G., Gao, Y., Han, J.: Approximating discrete probability distribution of image emotions by multi-modal features fusion. Transfer **1000**(1), 4669–4675 (2017)

14. Zheng, A., Xu, M., Luo, B., Zhou, Z., Li, C.: Class: Collaborative low-rank and sparse separation for moving object detection. Cogn. Comput. **9**(2), 180–193 (2017)

15. Zhong, G., Yan, S., Huang, K., Cai, Y., Dong, J.: Reducing and stretching deep convolutional activation features for accurate image classification. Cogn. Comput. **10**(1), 179–186 (2018)

Analysis of Land Cover Type Using Landsat-8 Data

V. Samuktha$^{(\boxtimes)}$, M. Sabeshnav, A. Krishna Sameera, J. Aravinth, and S. Veni

Department of Electronics and Communication Engineering, Amrita School
of Engineering, Amrita Vishwa Vidyapeetham, Coimbatore 641112, India
cb.en.u4ece18247@cb.students.amrita.edu, j_aravinth@cb.amrita.edu

Abstract. Classification of images attributes to categorizing of images
into various predefined groups. A particular image can be grouped into
several diverse classes. Examining and ordering the images manually
is a tiresome job particularly when they are abundant and therefore,
automating the entire process using image processing and computer
vision would be very efficient and useful. In this study, the Classifier
and Regression trees (CART) algorithm is used to create a classifier
model that classifies a region based on the feature specified. The Google
Earth Engine (GEE) platform is utilized to conduct the study. The Tier
1 USGS Landsat 8 surface reflectance dataset is employed and is sorted
according to the cloud cover. The features are then extracted and are
merged to obtain a feature collection. This input imagery is further sampled using particular bands from the Landsat imagery to get a renewed
feature collection of training data and the classifier model is trained using
the CART Algorithm. An accuracy assessment is further performed to
determine the exactness of the proposed model and the results are plotted using a confusion matrix. By applying the CART algorithm for image
classification, an accuracy of 83% is achieved which was found to be better than the existing results.

Keywords: USGS · Landsat 8 · GEE · CART algorithm · Reflectance

1 Introduction

Today, with the escalating necessity, capriciousness and the advancing demands
of technologies like artificial intelligence, disciplines such as machine learning,
and its subspaces have achieved enormous propulsion. The applications demand
tools, such as classifiers, which support an immense volume of data, interpret
them and derive features that are propitious. These classification methods aim in
categorizing the pixels of a digital image into various classes [1]. Usually, classification is implemented with the multi-spectral data and the spectral specimens
existing in the features of every pixel is utilized for grouping. The primary intention of the classification of images is to distinguish and mark the features in an

© IFIP International Federation for Information Processing 2021
Published by Springer Nature Switzerland AG 2021
V. Krishnamurthy et al. (Eds.): ICCIDS 2021, IFIP AICT 611, pp. 78–89, 2021.
https://doi.org/10.1007/978-3-030-92600-7_8

image and possibly plays the most essential role in digital image interpretation. Object classification is a complicated job and hence, image classification has a major part in the domain of computer vision. Classification is an art of labeling images into numerous categories. A particular picture can be grouped into several classes and the automation of the entire process of comparing and classifying images would surely make things easier than manual work. There are many real-time implementations which comprise computerized image design, extensive audio-visual databases, face recognition via social networks, and several other applications [2,3] which require classifiers to obtain high accuracy. Image Classification usually involves the following steps - Initially, we perform the Image pre-processing. Pre-processing of images generally involves the examination of the image, Resizing and Data Augmentation methods such as Gray scaling of images, Gaussian Blurring, Reflection, Equalization, Rotation, and Translation of images [4–7]. This step is succeeded by the extraction of features and training the model. This is a vital process where the analytical or machine learning techniques are applied in classifying the attractive attributes of the picture and extracting features that might be unprecedented over a distinct class, and this will improve the classification model in distinguishing the various classes. This method, known as model training, is the process where the classifier model learns the features from the dataset. These features are then applied to the classification stage for object detection. The detected objects are grouped into predetermined groups by employing suitable grouping techniques that correlate the image and the target patterns [5].

Many classification algorithms are used in image classification and these algorithms can be broadly categorized based on the type of classification techniques the algorithms apply. Supervised classification is predicated on the basis that the pixel specimens of an image, which represent specific classes, can be selected by the user. The image processing tools are then steered by these pixels to utilize these training sites for classifying all additional pixels in the picture [8]. Once every information class has been statistically characterized, the image is classified by performing a reflectance measurement for each pixel and selecting the signatures it relates the most. Classification algorithms and regression techniques are utilized by the supervised classifiers to develop predictive models. Few such algorithms which employ supervised classification are logical regression, random forests model, decision trees, support vector machine (SVM) classifiers, convolutional neural networks, Naive Bayes and k-nearest neighbours [9–11].

One of the most accurate and frequently applied supervised learning techniques are tree-based algorithms. Predictive modeling with higher efficiency, greater stability, and ease of interpretation are some of their advantages. Models such as decision trees, random forest, and gradient boosting are commonly applied in a variety of data science puzzles. Therefore, it's very effective to acquire the knowledge of these algorithms and implement them while modeling [3,12].

This study implements the decision tree algorithm, which is also known as the CART Algorithm is used to perform the classification of multispectral images. Landsat 8 dataset is acquired from USGS and is used for experimental analysis. It employs the concept of supervised learning and has a predefined objective. It is mostly applied in decision making which works on a non-linear basis and has a simple linear decision aspect. They are versatile for resolving any query at hand - classification or regression [13]. We have performed all our study on the platform of Google Earth Engine (GEE). It is a cloud-based web application that allows scientists, researchers and developers to discover corrections, chart trends and quantify variances on the Earth's surface by blending the multi-petabyte directory of satellite imagery and geospatial datasets with planetary-scale examination capacities. It provides a global-scale insight and allows ready-to-use datasets. It makes use of a simple, yet powerful API and presents convenient tools to the users. We have made use of this platform to build a classification model using the CART Algorithm. Our model can classify the features of a given area into vegetation, water bodies, fallow land, and other areas respectively.

The remaining sections are established as follows: Sect. 2 describes the various works related to this paper. Section 3 presents a detailed description of the methodology of the entire study. It explains the concepts of dataset selection, feature extraction, sampling the imagery, and training the classifier. Section 4 illustrates the results obtained during the process of the study and Sect. 5 provides the concluding remarks of the study.

2   Related Work

Classifying agricultural lands using remote sensing is a well-studied and implemented idea. Previous works have implemented the classification of cropland and fallow land using multispectral data from sensors with lower resolution like MODIS (250 m (bands 1–2) 500 m (bands 6) 1000 m (bands 8)). Zhuoting Wu and Prasad S. Thenkabail [14] have used MODIS data to classify the cropland. Kyle Pittman [15] also used MODIS data to map cropland in his work. Making use of the most advanced satellites will give us more accurate results. Using LANDSAT 8 (30 m (bands 1–5, 7) 60 m (bands 3–7) 15 m (bands 8)) products has an edge over other works which hasn't been used.

Jeena Elsa George, J Aravinth, and Veni S [16] calculated Top of Atmospheric reflectance manually for land surface temperature whereas we used the LANDSAT 8's TOA product which makes our solution more suitable to implement in real life applications.

Work based on cropland classification mainly uses algorithms like Support Vector Machine (SVM) for classification. Amit Kumar Bhasukala [17] in his work used the SVM algorithm in the classification process. Jhinzhong Kang [18] also used SVM to classify cropland and fallow land. This work uses the CART algorithm in the classification phase which gives us some advantages over previous works such as: not relying on data distribution, no overgrowth in the decision tree. The CART Algorithm has an edge over the other classification techniques and algorithms. Apart from its high accuracy, the CART Algorithm, which predominantly works based on Decision trees can perform multiclass classification.

It also provides the most model interpretability compared to the other algorithms which makes it more preferable. The features taken in the CART Algorithm have a non-linear relation which helps in not affecting the performance of decision trees. The fact that this algorithm can handle both numerical and categorical data which is most desirable for performing the classification. Both numerical and categorical data can be handled by the algorithm, making it desirable for performing the classification. M. Tugrul Yilmaz [19] in his work obtained an accuracy of just 70% using Decision tree classification.

From these studies it was observed that (i) lower resolution sensors were used for input data (ii) Manual computation of TOA has increased computational complexity (iii) Most of the classification models were highly dependent on SVM. To overcome these limitations, an attempt is made to incorporate the CART algorithm for obtaining accurate results in classifying satellite imagery and the right selection of the input data also plays an indispensable role. Using the proposed methodology, it is possible to attain a classification accuracy as high as 82.305% compared with the existing techniques used by various other authors.

3   Methodology

Figure 1 depicts the block diagram of the proposed system. It has the following stages: (i) Selection of dataset and Region of Interest (ROI) (ii) Image Pre-processing (iii) Feature extraction and (iv) Classification. As presented in Fig. 1, Tier 1 USGS Landsat-8 surface reflectance dataset acquired using the OLI/TIRS sensors is selected. As a step towards preprocessing, the datasets are sorted based on the cloud cover. The images comprise of five Visible and Near Infrared (VNIR) bands and two Short Wave Infrared (SWIR) bands. It is further treated to orthorectified surface reflectance. The region of interest encompasses a small town in the district of Erode along the banks of the river Kaveri. This region was particularly chosen, as this expanse includes a variety of different landforms making it optimal for training and classification.

The features are extracted by utilizing the point marker tool available in GEE. The feature extraction is performed based on a single label called land cover and is assigned different class numbers to the various features extracted. The features extracted in this study are vegetation, water, fallow and others which include urbanized areas, roads, etc. Bands B2, B3, B4, B5, B6, B7, and B10 are selected from the Landsat image for training and are then used to obtain a feature collection of training data by sampling the input imagery. The CART Algorithm, along with the extracted features is used to train the classifier model. The trained model is used to classify the image. For better visualization, a color palette is applied to display the images based on the corresponding color that has been assigned to it in the feature collection.

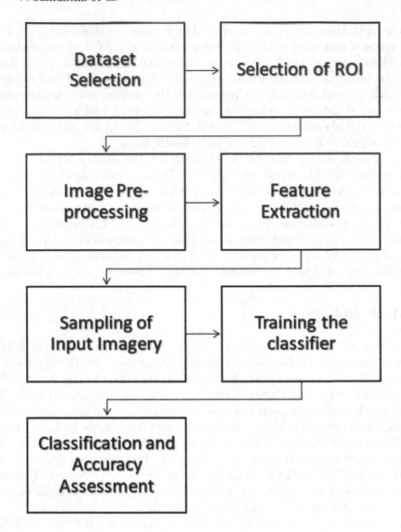

Fig. 1. The block diagram for the proposed methodology.

4   Results and Discussions

4.1   Dataset Selection

We have taken the Tier 1 USGS Landsat 8 dataset for our study. Landsat 8 dataset contains totally 11 bands namely coastal, blue, green, red, NIR, SWIR 1, SWIR 2, pan, Cirrus, TIRS 1, TIRS 2. Bands 1–7 and 9 has the resolution of 30 m and band 8 has the resolution of 15 m and bands 10 and 11 has 100 m as their resolution but resampled to 30 m. The data format of the Landsat 8 data is GeoTIFF with 16-bit pixel values. Surface reflectance images which are atmospherically corrected obtained from the Landsat Operational Land Imager

(OLI) and the Landsat Thermal Infra-Red Scanner (TIRS) sensors are included in this dataset [16].

These images comprise of five visible and near infrared (VNIR) bands and two short wave infrared (SWIR) bands. They are further treated to ortho-rectification of surface reflectance, and the two thermal infrared (TIR) bands are treated to temperature brightness ortho-rectification as shown in Fig. 2.

Managing Data Imbalancing. The arrangement of Land Cover (LC) classes is often imbalanced with some majority LC classes dominating upon minority classes. Although standard Machine Learning (ML) classifiers can deliver high accuracies for majority classes, they comprehensively fail to present reasonable accuracies for minority classes. This is essentially due to the class imbalance problem. In our study, a hybrid data balancing technique called the Partial Random Over-Sampling and Random Under-Sampling (PROSRUS), was applied to solve the class imbalance issue. Unlike many data balancing techniques which attempt to fully balance datasets, PROSRUS uses a partial balancing procedure with hundreds of fractions for a majority and minority classes for balancing datasets. For this, time-series of Landsat-8 along with several spectral indices was used within the Google Earth Engine (GEE) cloud platform. It was discerned that PROSRUS performed better than numerous other balancing methods and improved the precision of minority classes without affecting the overall classification accuracy.

The PROSRUS method blends the two well-known data-level balancing methods, ROS [20] and RUS [21]. ROS, a simple oversampling technique, randomly duplicates samples from minority class(es) to balance the distribution of classes. Balancing an original imbalanced dataset completely utilising this method could create overfitting of the classifier due to the duplication [22]. On the other hand, RUS randomly eliminates samples from the majority classes to fit the data distribution. The principal deficiency of a fully balancing dataset using RUS is that it may miss relevant data [23]. The hybrid method used in our paper not only takes the advantages of both ROS and RUS but also restricts their limitations by analysing 200 different fractions in the balancing design.

4.2   Selection of Region of Interest

We have taken our region of interest in the Erode district of Tamil Nadu, India as shown in Fig. 3. The region bounded by the four coordinates are considered as follows:
[77.6975208136198, 11.388427392274773],
[77.72799070802898, 11.388427392274773],
[77.72799070802898, 11.411985832912757],
[77.6975208136198, 11.411985832912757], We have particularly chosen these coordinates, as this expanse of the area encompasses several different landforms making it optimal for training and classification. After obtaining the images with the lowest cloud cover, we apply the false-color composite to the image and proceed with the subsequent steps.

Fig. 2. Original image data from study area

Fig. 3. The selected region of interest

4.3   Feature Extraction

The next step is to extract the features from our region of interest. Here, the point marker feature available in the Google Earth Engine (GEE) is utilized to perform this task. The features extracted are named accordingly and are imported as a feature collection and are given different properties. The features of vegetation zones, water bodies, fallow lands and other urban areas are extracted are shown in Fig. 4, 5, 6 and 7 respectively. The features are then merged under a single property, named land cover as shown in Fig. 8. After performing the feature extraction, around 621 elements are obtained in the feature collection. After retrieving the images with the least cloud cover, we use the false-color composite to the image and continue with the subsequent steps.

4.4   Sampling the Input Imagery

Bands B2, B3, B4, B5, B6, B7, and B10 are selected from the Landsat image for training the image and then use these bands for sampling the input images to get a feature collection of training data.

Fig. 4. Extracted features of vegetation

Fig. 5. Extracted features of water bodies

Fig. 6. Extracted features of fallow lands

Fig. 7. Extracted features of other areas

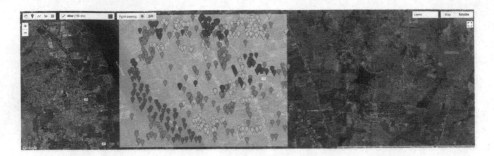

Fig. 8. Sampling the input imagery

Fig. 9. Trained classifier model

4.5 Training the Classifier

The classifier model is now trained using the extracted features by applying the CART algorithm. The trained model is used to classify the image. To have a better visualization, a color palette is incorporated to view the images based on the corresponding color that has been assigned in the feature collection as shown in Fig. 9. As the succeeding step, Land Use Land Cover Mask (LULC2010) is applied to classify the entire study area according to the colors specified in the color palette.

An accuracy assessment to determine the exactness of our model is conducted. Initially, a column of random uniforms was added to the dataset and the random function of GEE is used to split the datasets into testing and training datasets respectively. 70% of the dataset is used as the testing dataset and 30% as the training dataset. The model is trained using the training dataset and is tested with the testing dataset. The Confusion matrix, a tool usually used to estimate the performance of machine learning problems, is adopted to plot the results as an error matrix, to determine the accuracy of the model. The confusion matrix obtained by this classifier is presented in Table 1.

The image is classified based on the supervised classification method and to calculate the efficiency of the classification, the dataset collection has been split into a test set and training set where the training dataset records for 70% of the images and is used to train the classifier. The rest of the images were taken as

Table 1. The confusion matrix of classified features upon the specific ROI for accuracy assessment

Class	Accuracy	Precision	Recall	F1Score
Vegetation	84.77%	0.95	0.79	0.86
Waterbodies	100%	1	1	1
Fallow land	87.24%	0.63	0.78	0.69
Other	92.59%	0.68	0.94	0.79

the testing dataset. To compute the accuracy of the trained model, it is executed with the testing dataset. Then it classifies the pixels of the test set based on the features that it has been trained to classify. The overall accuracy obtained in classifying the four classes of features is 82.305%.

José M. Peña-Barragán, Moffatt K. Ngugi, Richard E. Plant, Johan Six [24] have carried out the assessment of crops by applying Object based Image Analysis (OBIA) along with various vegetation indices and crop phenology and obtained an overall accuracy of 79% and our work proved to be more robust in classifying the features in terms of accuracies, by applying the same classification model Decision Tree as they did.

5 Conclusion and Future Works

An improved method of multi-spectral image classification was attempted by using the CART algorithm, which leads to promising results as compared to the existing techniques. The Tier 1 USGS Landsat 8 surface reflectance dataset, which is a multispectral dataset, is employed and is ordered according to the cloud cover. The features were extracted and merged to achieve a collection of features. This input imagery is additionally sampled using the particular bands from the Landsat imagery to obtain a renewed feature collection of training data, and the classifier model is trained using the CART algorithm. An accuracy assessment is further performed to determine the exactness of the model developed and an overall accuracy of 82.305% was achieved. From this study, it is observed that the use of decision tree based algorithms enhanced the performance of the classification model with increased accuracies. This model can be employed in the fallow land classification as a future work.

References

1. Abburu, S., Golla, S.B.: Satellite image classification methods and techniques: a review. Int. J. Comput. Appl. **119**(8), 20–25 (2015)
2. Basavanhally, A.N., et al.: Computerized image-based detection and grading of lymphocytic infiltration in HER2+ breast cancer histopathology. IEEE Trans. Biomed. Eng. **57**(3), 642–653 (2009)

3. Choi, K., Toh, K.-A., Byun, H.: Incremental face recognition for large-scale social network services. Pattern Recogn. **45**(8), 2868–2883 (2012)
4. Sathya, S., Joshi, S., Padmavathi, S.: Classification of breast cancer dataset by different classification algorithms. In: 2017 4th International Conference on Advanced Computing and Communication Systems (ICACCS), pp. 1–4. IEEE (2017)
5. Kamavisdar, P., Saluja, S., Agrawal, S.: A survey on image classification approaches and techniques. Int. J. Adv. Res. Comput. Commun. Eng. **2**(1), 1005–1009 (2013)
6. Prabhakar, T.V.N., Geetha, P.: Two-dimensional empirical wavelet transform based supervised hyperspectral image classification. ISPRS J. Photogramm. Remote. Sens. **133**, 37–45 (2017)
7. Kavitha Balakrishnan, S.V., Soman, K.P.: Spatial preprocessing for improved sparsity based hyperspectral image classification. Int. J. Eng. Res. Technol. **01**, 1–5 (2012)
8. Gómez, C., White, J.C., Wulder, M.A.: Optical remotely sensed time series data for land cover classification: a review. ISPRS J. Photogramm. Remote. Sens. **116**, 55–72 (2016)
9. Bhavsar, H., Ganatra, A.: A comparative study of training algorithms for supervised machine learning. Int. J. Soft Comput. Eng. (IJSCE) **2**(4), 2231–2307 (2012)
10. Rithin Paul Reddy, K., Srija, S.S., Karthi, R., Geetha, P.: Evaluation of water body extraction from satellite images using open-source tools. In: Thampi, S.M., et al. (eds.) Intelligent Systems, Technologies and Applications. AISC, vol. 910, pp. 129–140. Springer, Singapore (2020). https://doi.org/10.1007/978-981-13-6095-4_10
11. Saravanamurugan, S., Thiyagu, S., Sakthivel, N.R., Nair, B.B.: Chatter prediction in boring process using machine learning technique. Int. J. Manuf. Res. **12**(4), 405–422 (2017)
12. Singh, S., Gupta, P.: Comparative study ID3, cart and C4. 5 decision tree algorithm: a survey. Int. J. Adv. Inf. Sci. Technol. (IJAIST) **27**(27), 97–103 (2014)
13. Thambi, S.V., Sreekumar, K.T., Santhosh Kumar, C., Reghu Raj, P.C.: Random forest algorithm for improving the performance of speech/non-speech detection. In: 2014 First International Conference on Computational Systems and Communications (ICCSC), pp. 28–32. IEEE (2014)
14. Wu, Z., et al.: Seasonal cultivated and fallow cropland mapping using MODIS-based automated cropland classification algorithm. J. Appl. Remote Sens. **8**(1), 083685 (2014)
15. Pittman, K., Hansen, M.C., Becker-Reshef, I., Potapov, P.V., Justice, C.O.: Estimating global cropland extent with multi-year MODIS data. Remote Sens. **2**(7), 1844–1863 (2010)
16. George, J.E., Aravinth, J., Veni, S.: Detection of pollution content in an urban area using Landsat 8 data. In: 2017 International Conference on Advances in Computing, Communications and Informatics (ICACCI), pp. 184–190. IEEE (2017)
17. Basukala, A.K., Oldenburg, C., Schellberg, J., Sultanov, M., Dubovyk, O.: Towards improved land use mapping of irrigated croplands: performance assessment of different image classification algorithms and approaches. Eur. J. Remote Sens. **50**(1), 187–201 (2017)
18. Kang, J., Zhang, H., Yang, H., Zhang, L.: Support vector machine classification of crop lands using sentinel-2 imagery. In: 2018 7th International Conference on Agro-Geoinformatics (Agro-Geoinformatics), pp. 1–6. IEEE (2018)
19. Yilmaz, M.T., Hunt, E.R., Jr., Goins, L.D., Ustin, S.L., Vanderbilt, V.C., Jackson, T.J.: Vegetation water content during SMEX04 from ground data and Landsat 5 thematic mapper imagery. Remote Sens. Environ. **112**(2), 350–362 (2008)

20. Chawla, N.V.: Data mining for imbalanced datasets: an overview. In: Maimon, O., Rokach, L. (eds.) Data Mining and Knowledge Discovery Handbook, pp. 875–886. Springer, Boston (2009). https://doi.org/10.1007/978-0-387-09823-4_45
21. Batista, G.E., Prati, R.C., Monard, M.C.: A study of the behavior of several methods for balancing machine learning training data. ACM SIGKDD Explor. Newslett. **6**(1), 20–29 (2004)
22. Santos, M.S., Soares, J.P., Abreu, P.H., Araujo, H., Santos, J.: Cross-validation for imbalanced datasets: avoiding overoptimistic and overfitting approaches [research frontier]. IEEE Comput. Intell. Mag. **13**(4), 59–76 (2018)
23. Haixiang, G., Yijing, L., Shang, J., Mingyun, G., Yuanyue, H., Bing, G.: Learning from class-imbalanced data: review of methods and applications. Expert Syst. Appl. **73**, 220–239 (2017)
24. Peña-Barragán, J.M., Ngugi, M.K., Plant, R.E., Six, J.: Object-based crop identification using multiple vegetation indices, textural features and crop phenology. Remote Sens. Environ. **115**(6), 1301–1316 (2011)

Rule Based Combined Tagger for Marathi Text

Kalpana B. Khandale[(⊠)] and C. Namrata Mahender

Department of Computer Science and IT, Dr. Babasaheb Marathwada University, Aurangabad,
Maharashtra, India
cnamrata.csit@bamu.ac.in

Abstract. Part of speech (POS) tagging is most necessary concept in the natural language processing categorized each word in the corpus. This paper focus on the development of Marathi part of speech tagger using the N-gram models. For this we have designed rules for the development of the POS tagger. These rules are framed on the basis of the Marathi grammar. The corpus is of 635 Marathi sentences written in considering variations, for better evaluation of the POS tagger. Total 5715 words has been tagged from in which 1918 words unigram tagger and 106 words are tagged by the bigram tagger. The overall accuracy of POS tagger is 79.34%.

Keywords: Annotated corpus · Unigram · Bigram · POS tagging

1 Introduction

Around 90 million people speak the Marathi language. Large amount of information available in Marathi in electronic form and this information is not in standard format and the uniformity of the text is the main problem. Many tools available for the normalization of the text but those tools are not supported to the Marathi up to the mark. Therefore, neither the corpus nor the processing tools available for the Marathi language. The tools available in the other language are not adequate or they are limited to some domain specific. Almost all the application of natural language processing there is need the part of speech tagger for identifying the grammatical tag of each word. Marathi is the verb final language relatively free order language. Word order is typological property of any language hence the study of syntax word order is not complete when there is no deliberation or discussion on the word order of any language. The order of Marathi language is SOV (Subject Object Verb). In Marathi many word do not follow the strict order in the sentences. For example,

© IFIP International Federation for Information Processing 2021
Published by Springer Nature Switzerland AG 2021
V. Krishnamurthy et al. (Eds.): ICCIDS 2021, IFIP AICT 611, pp. 90–101, 2021.
https://doi.org/10.1007/978-3-030-92600-7_9

Table 1. Interchangeable words in sentence.

Sr. No.	Marathi Sentence	English Sentence
1.	मनोज रविला उद्या भेटवस्तू देईल.	
2.	मनोज उद्या भेटवस्तू रविला देईल.	
3.	मनोज उद्या रविला भेटवस्तू देईल.	
4.	रविला उद्या मनोज भेटवस्तू देईल.	Tomorrow Manoj will
5.	उद्या रविला मनोज भेटवस्तू देईल.	give a gift to Ravi.
6.	उद्या रविला मनोज भेटवस्तू देईल.	

The above Table 1 depicts the same meaning but the order of the words is inter-changeable. The meaning of the entire sentence is "Tomorrow Manoj will give a gift to Ravi." It is so simple in English but in Marathi sometimes when word order has been change but the meaning of the sentence has not been change. But on the other hand Marathi has such unit which does not change that order like verb followed by the auxiliary verb, postpositions always follow the noun, adjectives come before the noun, etc. On the basis of the transformational grammar of Marathi language, noun phrase looks like the different meanings as described in the following example (Table 2)

Table 2. Transformational grammar based noun phrase.

Sr. No.	Marathi Words	English Words
1	मंदिर	A temple
2	मंदीरासाठी	For temple
3	मंदीरावर	On the temple
4	मंदीराजवळ	Near the temple
5	मंदिरात	In the temple
6	मंदीरामध्ये	In the temple
7	मंदीराचे	Of the temple
8	मंदीरामधील	In the temple
9	मंदीरासमोरील	In front of the temple
19	मंदीरामागे	Behind the temple

From the above table the context of the 'मंदीर', if different types of suffixes and postpositions add into the word then the meaning has been change. From the above example, we can understand the importance of the Marathi POS tagger. Because when the suffixes and the postpositions are added into the word the tagging of the word will be change.

1.1 Part of Speech Tagger

In language, words have a number of different meanings. Contemplate the formal and functional distinction of words and they are grouped into the different word classes called as the part of speech tagger. There are different techniques are available for finding out the word level ambiguity, and it is based on the context of the words in the text under consideration with the respective words with POS. The work of the POS tagging for Marathi is based on the different types of rules. The rules of the POS tagging for Marathi is based on the categories of words belong to. Marathi is one of the Prakrit languages which are developed from Sanskrit. Marathi is the free ordered language and also a verb final language. There are some types of part of speech tagging described as below.

1.1.1 Rule Based Part of Speech Tagger

One of the most useful and oldest techniques of word tagging is the rule based POS tagging in which we can use the handwritten rules for tagging the word. If word has more than one feasible tag, then this rules are useful to identifying the correct tag of the word. For Marathi is it necessary because the sentences below have the more complexity for identifying the correct tag for the word. For example,

1) तू झाड लाव.
2) तू घर झाड.

In these examples the 'झाड' has two meaning into two different sentences. In the first sentence the word 'झाड' is tagged as noun and in the second sentence, is it tagged as verb. The advantage of rule based tagger is that knowledge driven tagger and its built manually.

1.1.2 Stochastic Tagger

This tagger is also called as the statistic POS tagger. This is included the frequency and probabilities of words. Here we can use the word frequency and encountered the most frequent words which are present in the training set. On the other hand, the probability of tag also calculated. It requires that training corpus of tagged sentences. For this tagger if word is not present in the corpus it is not calculating the probability of that word.

1.1.3 Transformation Based Tagging

It is also called as Brill tagging. This allows linguistic knowledge in the readable form and transforms it into one state to another state by using this transformation based rules. The advantage of this tagging it reduced the complexities of tagging because in this tagger there is entwine of the machine learning and hand written or human generated rules are there.

2   Literature Survey

Table 3. Previous study on POS tagger

Author, Year and Language	Methods/Technique	Dataset	Issues/limitation	Result
AlKhwiter, W., & Al-Twairesh, N. 2021 (Arabic) [1]	-Conditional Random Fields and Bidirectional Long Short-Term Memory (Bi-LSTM) models	-Total 7750 tweets from three different types of tweets -Tweets with a length of less than seven words -Spam tweets -Tweets that were written in dialectal Arabic	-The tagger was not able to tag twitter-specific items correctly -There was need a good tokenizer as tool in NLP for Arabic	-The overall accuracy of 'Mixed' dataset was 96.5% -'MSA' and 'GLF' datasets achieve an accuracy of 95.6% and 95% respectively
Bacon, G. 2020 (Latin) [2]	-Unidirectional LSTM -Hyperparameter optimization	-The EvaLatin 2020 shared task dataset -Containing 14,399 sentences with of 259,645 words	-Not specified	-The overall performance of the system was above 95.3%
Yousif, J., & Al-Risi, M. 2019 (Arabic) [3]	-SVM -Evaluation Measures	-Which contains 131 or 77 tags -Training and testing the SVM models starting with 1K, and the maximum size is 156K	-More research in the direction of analyzing and categorizing the Arabic tags and tag sets -Needed to build a standard Arabic corpus for NLP applications	-Not mention specifically
Kurniawan, K., & Aji, A. F. 2018 (Indonesian) [4]	-Bidirectional LSTM (biLSTM) -Scikit-learn library -Majority tag (MAJOR) -Memorization (MEMO) -CRF	-The corpus contains 10K sentences and 250K tokens that are tagged manually	-Not specified	-The POS tagger demonstrated an accuracy of 69%
Nita. V. Patil 2018 (Marathi) [4]	-N-gram -HMM used the Viterbi algorithm	-From the domain of news stories they have used 15,000 sentences	-POS tagging is challenging task for Marathi	-The overall accuracy of the system is 86.61%
Ajees A P, Sumam Mary Idicula 2018 (Malayalam) [5]	-Statical approach -CRF Model -Maximum entropy Markov models	-They have used to 23K words for training and 5.7K words for testing	-Not specified	-The overall accuracy of the system is 91.2%

(continued)

Table 3. (*continued*)

Author, Year and Language	Methods/Technique	Dataset	Issues/limitation	Result
-Mittal, S., Sethi, N. S., & Sharma, S. K. 2014 (Punjabi) [6]	-Bigram model	-They have collected 2400 sentences from 10,000 words randomly	-The performance of the n-gram based method was not so accurate because when encountered the unknown words like foreign language words	-The accuracy was 92.12%
Joshi, N., Darbari, H., Mathur, I. 2013 (Hindi) [7]	-HMM -Evaluation matrix	-From the domain of the tourism 15,200 sentences with 3, 58,288 words for trained the system	-They have not concentrate on adding more tags the classification of the text	-The accuracy of the system is 92.13%
Singh, J., Joshi, N., &Mathur, I 2013 (Marathi) [8]	-Statistical approach -Trigram model -Accuracy	-2000 sentences with 48,635 words have been used for the trained the system	-The morphological complexity of the Marathi is hard to identify the morphemes	-The accuracy was 91.63%
JyotiSingh,ItiMathur&Nisheeth Joshi 2013 (Marathi) [9]	-Statistical Approach -Unigram -Bigram -Trigram -HMM methods	-They have used 1000 sentences with 25744 -Words for trained the system	-The data they have been used was less	-The overall system result was 93.82%
Dhanalakshmi, V., Shivapratap, G., &SomanKp, R. S. 2009 (Tamil) [10]	-SVM tools -Non-linear SVM using Linear programming	-25,000 Sentences -10,000 sentences have used for the testing	-Corpus is not available in Tamil - Nouns get inflected for number and cases. Verbs get inflected for tense, person, number, gender suffixes	-The overall accuracy 95.63%
Patel, C., & Gali, K. 2008 (Gujrati) [11]	-Machine learning approach -Conditional Random Fields (CRF) model	-Out of 600 sentences 10,000 words they have used for training and 5,000 words used for the testing	-The CRF used for the features and the probabilities of the tag because of the flexible nature of the language	-The accuracy of the system was 92%
Asif Ekbal, Sivaji Bandyopadhyay 2008 (Bengali) [12]	-Hidden Morkov Model (HMM) -Support Vector Machine (SVM)	-They have used the 72,341 words for trained the system	-The data they have used were very less	-The overall result was 91.23%

(*continued*)

Table 3. (*continued*)

Author, Year and Language	Methods/Technique	Dataset	Issues/limitation	Result
Singh, T. D., & Bandyopadhyay, S. 2008 (Manipuri) [13]	-Machine learning approach -Accuracy	-Out of 3784 sentences 10917 unique words have been used for the training purpose	-Disambiguation scheme is necessary for the Noun-Adjective ambiguity	-The overall accuracy was 69%
Dalal, A., Nagaraj, K., Sawant, U., & Shelke, S. 2006 (Hindi) [14]	-Maximum Entropy Markov Model (MEMM)	-They have used NLPAI-ML 2006 contest which have consisting of around 35000 words	-The features which has language specific that would be improve the performance system particularly in case of chunking	-The total accuracy of 88.4% for POS tagging and 86.45% for chunking
Coden, A. R., Pakhomov, S. V., Ando, R. K., Duffy, P. H., & Chute, C. G. 2005 (English) [15]	-Trigram models on MED data	The Penn Treebank-2 corpus -The GENIA corpus -MED corpus of clinical notes	-Not specified	-The accuracy of the system was 92% accuracy from 87% in our studies

3 Proposed System

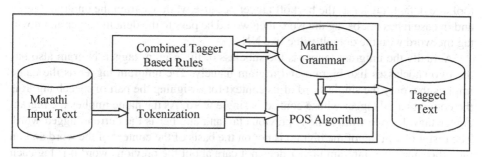

Fig. 1. Architecture of POS tagger

3.1 Input Text

As specific corpus of Marathi text or dataset is not available, 907 sentences were manually created, with many variations to sense the difficulties while generating tags using the proposed POS tagger (Fig. 1, Table 3).

3.2 Tokenization

Tokenization is the process of splitting each word from the input text. In Marathi the words have separated with the white space and the punctuation marks. With the help of this we can easily tokenize the sentence into separate tokens.

3.3 Marathi Grammar

The tag to the word are assigned on the basis of the Marathi grammar rules. Marathi is the augmented language. It has eight types of part of speech (Naam (noun), Sarvnam (pronoun), Kriyapad (verb), Visheshan (Adjectives), Shabdyogi Avyay (Postposition), Kriya Visheshan Avyay (Adverb), Ubhayanvayi Avyay (Conjunction), and Kevalprayogi Avyay (Interjection)).

3.4 Combined Tagger Based Rules

Our aim to develop the part of speech tagger for Marathi and for that we have used here the combined tagger to train the data.

3.4.1 Combined Tagger

Combination of taggers is one of the main features in natural language processing. The importance of the combined tagger is that when one tagger is unable to tag the word then it would be passed another one is called as sequential backoff tagging. In our case we took the default tagger as the backoff tagger because while we train the unigram tagger and in case it has not be tag the word then would be pass to the default tagger and it will tag the word with the default value like NN.

Actually the unigram tagger is the subclass of the n-gram tagger. N-gram also has the two subclasses like bigram and trigram namely. The unigram tagger as the name implies that only the single word of its context for assigning the part of speech tag. It is the context based tagger whose context is single word. As the name implies the bigram tagger tags the two words one is previous tag and another is the current tagged word. The result of tagging of the bigram is not on the basis of the context of the word and on the other hand the unigram tagger does not care about the previous word it is tag each current word. So while we combine both the tagger the result is better. Rules framed for the POS tagging, some of them given as below (Table 4),

Table 4. Sample rule for POS tagging.

Rules For Noun	If suffixes or postpositions precedes the word: Then tag as the NN
Rules for Verb	If NN or JJ or RB in sentence: Then tag as VM Else NN or JJ or VM in sentence: Then tag as VAUX
Rule for Adjective	If preceding word of noun present in sentence: Then tag as JJ

Algorithm for part of speech tagger:

Step 1: Take Marathi text as input.
Step 2: Create list of tokenized words from sentences of input text.
Step 3: Apply rule based POS tagger on the tokenized text.
Step 4: If the tag is not correct repeat the Step 3.
Step 5: Stop.

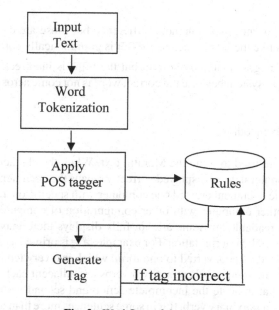

Fig. 2. Workflow of the system

Main issues encountered while designing a rule based POS tagger

1. Some issues are encounters while tagging text of Marathi language because it has words like (बोलता-बोलता, दुख:) these words include the hyphen (-) and colon (:) but no other punctuation mark. But hyphen is come across into the sentences it is treated as the individual token. And the 'बोलता' and 'बोलता' also consider two different tokens in the sentence.
2. While doing the segmentation with the full stop at the end like (पु.ल.देशपांडे) sustain the segmentation ambiguity. The full stop denotes the end of the sentence as like English.
3. Another issue, the word in Marathi like 'नवी मुंबई', it is proper noun but during the tokenization it is considered as two different tokens as 'नवी' and 'मुंबई'.
4. Variation in spelling: There are word containing four vowels in Marathi described in the below table (Table 5),

Table 5. Vowels in Marathi

Vowel in Marathi	Vowel Sign
इ(i)	(िँoo)
ई(ī)	(ooी)
उ(u)	(ोॢ)
ऊ (ū)	(ोॢ)

All the vowels do not make phonetic difference but those are differing in writing style and spellings like the words such as 'विराज'is grammatically correct but 'वीराज'is incorrect or 'राणी' is grammatically correct but the 'रानि'is incorrect. These affect the part of speech tagger system because the correct word is not come across into the training dataset (Fig. 2)

5. Issue related to encoding:

Various fonts are used to write the Marathi text. When the character of language is opened in the computer it is not displayed correctly means the document is unreadable or unusable. When the document created one computer with specific operating system may not displayed another computer with other configuration of computer. In these cases, some character is readable but some are not fully displays there may be some hollow circles or some symbolic interpretation. For example, केंद्रis printed as केंद. Thus all font types has to be traced and converted to one form, which is a very tedious job.

While tagging the sentence there are two verbs are adjacent each-other. First word act as main verb that indicate the incomplete action and second verb that indicate the complete action act as auxiliary verb. But in some sentences more than one verb occurred. For example, 'मोहन राधाला कुलफी घेऊन देत होता.' In this type of example, the verb like 'घेऊन'and 'देत' are adjacent and one auxiliary verb that is 'होता'.

4 Result

In the following Fig. 3 and 4, POS tagger is shown with the tagged data on the training data and the testing data. Total 635 tagged sentences are considered for the development of the part of speech tagger. For training 90% (571 tagged data) of the tagged sentences and for testing 10% (126 tagged data) sentences are used from the total sentences. The below Fig. 3 depict the result of the combined tagger and we mainly concentrate on the unigram and bigram tagger, Total tagged words are 5715 and the size of the unigram tagger is 1918 words are tagged with unigram trained tagger and remaining untagged by the unigram is passed to the bigram tagger and bigram tagger is has successfully tagged 106 words. The table below shows some sentences and our trained tagger results. The overall result of the POS tagger for Marathi is 79.34%. The performance can be enhanced by increasing the training corpus.

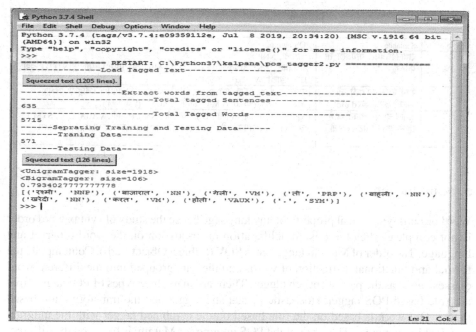

Fig. 3. Result of the Marathi POS tagger

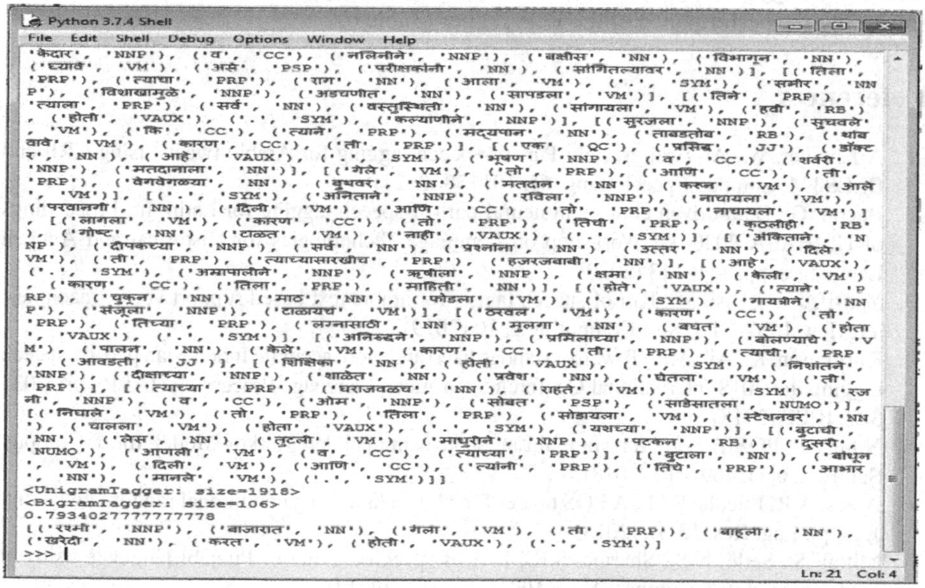

Fig. 4. Sample of trained POS tagger

Table 6. Sample dataset with result

Sr. No.	Sentence	POS tagging
1.	राधाने चित्रपट बघितला, ती फार खुश झाली.	('राधाने','NNP') ,('चित्रपट','NN','बघितला','VM'),(',',',','SYM'),('ती','PRP'),('फार','INTF'),('खुश', 'JJ'),('झाली','VM'),('.','SYM')
2.	रश्मी बाजारात गेली , ती बाहुली खरेदी करत होती.	(रश्मी, 'NNP'), ('बाजारात', 'NN'), ('गेली', 'VM'), ('ती', 'PRP'), ('बाहुली', 'NN'), ('खरेदी', 'NN'), ('करत, 'VM'), ('होती', 'VAUX'), ('.', 'SYM')
3.	सुनील चांगला व्यक्ती आहे , तो सगळ्यांची मदत करतो.	('सुनील','NNP'), ('चांगला','JJ'), ('व्यक्ती','NN'), ('आहे','VAUX'),(' ,','SYM') ('तो','PRP'), ('सगळ्यांची','NN'),(' मदत','NN'),(' करतो','VM'),('.','SYM')

5   Conclusion

Word order is typological property of any language hence the study of syntax word order is not complete when there is no deliberation or discussion on the word order of any language. The order of Marathi language is SOV (Subject Object Verb). Contemplate the formal and functional distinction of words and they are grouped into the different word classes called as the part of speech tagger. There are main three types of POS tagger that are rule based POS tagger, stochastic or statistic tagger and the transformation based tagger. Our work is based on the rule based on the combined tagger with the unigram and the bigram tagger. The work of the POS tagging for Marathi is based on the different types of rules. Total 5715 words tagged during the training process and the overall result of the POS tagger is 79.34%. In future we will be adding the more data for training for enhancing the performance of the POS tagger (Table 6).

References

1. AlKhwiter, W., Al-Twairesh, N.: Part-of-speech tagging for Arabic tweets using CRF and Bi-LSTM. Comput. Speech Lang. **65**, 101138 (2021)
2. Bacon, G.: Data-driven choices in neural part-of-speech tagging for Latin. In: Proceedings of LT4HALA 2020–1st Workshop on Language Technologies for Historical and Ancient Languages, pp. 111–113 (2020)
3. Yousif, J., Al-Risi, M.: Part of speech tagger for arabic text based support vector machines: a review. ICTACT J. Soft Computing **1**, 10 (2019)
4. Kurniawan, K., Aji, A.F.: Toward a standardized and more accurate Indonesian part-of-speech tagging. In: 2018 International Conference on Asian Language Processing (IALP), pp. 303–307. IEEE (2018)
5. Nita, V.P.: POS Tagging for Marathi Language using Hidden Markov Model. Int. J. Comput. Sci. Eng. **6**(1), 409–412 (2018)
6. Ajees, A.P., Idicula, S.M.: A POS tagger for Malayalam using conditional random fields. Int. J. Appl. Eng. Res. **13**(3), (2018)
7. Mittal, S., Sethi, N.S., Sharma, S.K.: Part of speech tagging of Punjabi language using N gram model. Int. J. Comput. Appl. **100**(19), 19–23 (2014)
8. Joshi, N., Darbari, H., Mathur, I.: HMM based POS tagger for Hindi. In: Proceeding of 2013 International Conference on Artificial Intelligence, Soft Computing (AISC-2013) (2013)
9. Singh, J., Joshi, N., Mathur, I.: Part of speech tagging of Marathi text using trigram method. In: Communications and Informatics (ICACCI), 2013 International Conference on, pp. 1554–1559. IEEE (2013)

10. Dhanalakshmi, V., Shivapratap, G., SomanKp, R.S.: Tamil POS tagging using linear programming (2009)
11. Patel, C., Gali, K.: Part-of-speech tagging for Gujarati using conditional random fields. In: Proceedings of the IJCNLP-08 Workshop on NLP for Less Privileged Languages (2008)
12. Ekbal, A., Bandyopadhyay, S.: Web-based Bengali news corpus for lexicon development and POS tagging. Polibits **37**, 21–30 (2008)
13. Singh, T.D., Bandyopadhyay, S.: Morphology driven Manipuri POS tagger. In: Proceedings of the IJCNLP-08 Workshop on NLP for Less Privileged Languages (2008)
14. Dalal, A., Nagaraj, K., Sawant, U., Shelke, S.: Hindi part-of-speech tagging and chunking: a maximum entropy approach. In: Proceedings of the NLPAI Machine Learning Contest (2006)
15. Coden, A.R., Pakhomov, S.V., Ando, R.K., Duffy, P.H., Chute, C.G.: Domain-specific language models and lexicons for tagging. J. Biomed. Inform. **38**(6), 422–430 (2005)

Evaluating Candidate Answers Based on Derivative Lexical Similarity and Space Padding for the Arabic Language

Samah Ali Al-azani[✉] and C. Namrata Mahender

Department C.S. and I.T, Dr. Babasaheb Ambedkar Marathawada University, Aurangabad, Maharashtra, India

Abstract. Character difference represents one of the most common problems that can be occurred when students try to answer questions of fill in the gaps or one-word answer that is needed mostly to one word as the answer. To improve the evolution of the student answer using Hamming distance, we proposed Hamming model tried to solve the drawbacks of the standard Hamming model by applying the stemming approach to achieve derivative lexical similarity and applying the space padding to deal with unequal lengths of the texts.

Keywords: Hamming · Lexical similarity · Derivatives · Questions answering system

1 Introduction

1.1 Question Answering System

Questions Answering system is usually a challenging task, is a software engineering discipline inside the fields of data recovery and characteristic language handling (NLP), which is involved in building frameworks that consequently answer addresses presented by people in a characteristic language. The fundamental purpose of the questionnaire is to provide direct answers to users' questions in the native tongue [1, 2]. This has the incredible position of aiding clients find the answers they want without having to search for large reservoirs of acquaintance. Not at all like web index, for example, Google and yahoo, which permit the client to recover web pages or on the other hand reports that less sensitive to the specified keywords while leaving the function of extracting hyperlinks inactive and finding required verses from clients, QA programs provide the user with the most relevant details in answering their inquiries in the section or at the sentence level [3].

1.2 Challenges of Arabic QAS

Arabic is a Semantic language [4]. It has a population of over 422 million people worldwide. The first language of the Arab and the official language of the United Nations

© IFIP International Federation for Information Processing 2021
Published by Springer Nature Switzerland AG 2021
V. Krishnamurthy et al. (Eds.): ICCIDS 2021, IFIP AICT 611, pp. 102–112, 2021.
https://doi.org/10.1007/978-3-030-92600-7_10

[5]. It is the third most important international language after English and French. The Arabic language has a very rich combination of special features that the computer is difficult to perform [6]. This advantage has created many challenges that researchers have to deal with differently. This section looks at many of these challenges:

1. No capital letters.
2. Lack of linguistic resources.
3. Optional short vowels.
4. Free order senetences.
5. Arabic an inflectional language (word = root + affixes (prefix, infex, suffix).

2 Text Similarity Algorithms

A similarity measure is a work that allocates an actual number between 0 and 1 to the whole document. Zero esteem implies that the reports are different completely, whereas one demonstrates that the reports are identical essentially. Vector-based models have been utilized for computing the similarity in files, customarily. The various features presented in the files are given to by vector-based models. Text similarity measurement shows a development role that connected research and all applications in many tasks such as text classification, information retrieval, clustering document, and question answering system, text summarization, detection and correction, machine translation.

2.1 String-Based Similarity

String similarity is managed operation on strings sequence and structure of characters. A metric that is applied to measure the distance between the text strings is called a string metric. It is applied to matching strings. There are two types of string similarity functions, character-based similarity functions, and term-based similarity functions.

2.1.1 Character-Based Similarity

The character-based similarity is additionally known as the sequence-based or edit distance (ED) dimension. proceeds two characters strings and after that figure the edit distance (counting addition, cancellation, and substitution) between these strings., edit distance is broadly utilized for string coordinating estimation to deal with the current data irregularity information inequality [7]. There are some algorithms in this approach illustrative in Fig. 1.

2.1.2 Term-Based Similarity

This type is known as token-based since it displays each string as a bunch of tokens. the comparability between these strings can be assessed by controlling sets of tokens, such as words. The most thought behind this approach is to perform two string similarity estimation based on common tokens, compare to its token sets [8]. Term based similarity is best utilized on same length token evaluation. Are a few illustrations of these strategies as Cosine similarity [9], Jaccard similarity [10], Manhattan distance [11], and Euclidean distance [12], Dice's coefficient [13].

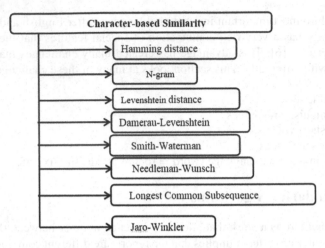

Fig. 1. Examples of character-based approach.

2.2 Corpus-Based Similarity

The corpus-based similarity is applied on a semantic likeness degree, which decides the likeness between the words based on the data picked up from corpora. Text corpus comprises of an expansive and organized set of writings, mostly corpus-based similarity are based on idea established resources, like Wikipedia. There are some types of corpus-based similarity approaches like Semantic Analysis (LSA) [9], Normalized Google Distance (NGD), Explicit Semantic Analysis (ESA) [14].

2.3 Knowledge-Based Similarity

A semantic likeness measure that employs data from semantic systems to distinguish the degree of word closeness is known as a knowledge-based likeness measure [15]. Knowledge-based likeness comprises semantic similitude and semantic relatedness. Those concepts have been energetically examined among around the world analysts. Knowledge-based similarity measures divided into the two group's measurements are semantic similarity and semantic relatedness. Measures of semantic similarity have been the focus on words and concepts [16]. The semantic approach employs an express representation of information, for example, the interconnection of realities, the implications of words, and rules to depict conclusions on particular spaces. The pattern of information illustration, by and large, incorporates the rules of eventuality, coherent recommendations, and arranges semantics such as scientific categorization and philosophy. The conspicuousness of word-to-word similitude measurements is due to the asset accessibility and particularly encodes relations between words and ideas (e.g. WordNet), the knowledge-based closeness approach that employments WordNet metaphysics can be categorized into three measures as Fig. 2. Semantic relatedness alludes to human judgments of the degree to which a given combination of concepts is related. Semantic relatedness and semantic likeness are two isolated ideas. Semantic relatedness could be a more common idea of the relatedness of concepts, whereas similitude could be an extraordinary case of relatedness that's tied to the resemblance of the concepts.

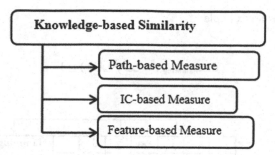

Fig. 2. Knowledge-based similarity approach

2.4 Hybrid-Based Similarity

The goal of this path is to incorporate the already portrayed approaches, counting string-based, corpus-based, and knowledge-based similitude to reach better results; a metric by receiving their preferences (Table 1).

Table 1. Common examples of hybrid similarity

No	Name of authors	Examples of hybrid metrics
1	Monge and Elkan [17]	Assume a recursive matching scheme to compare two long strings
2	Wang et al. [17]	Employed fuzzy matching between tokens
3	Cohen [18]	Apply hybrid metric "soft" TF-IDF similarity uses the Jaro-Winkler
4	Lin [19]	Proposed a novel linked data (LD) based on a hybrid semantic similarity measure, called TF-IDF (LD)
5	Al-Hasan [20]	Proposed a new presumed Ontology-based Semantic Similarity (IOBSS) measure to evaluate semantic similarity
6	Atoum and Otoom [21]	Develop a novel hybrid on quantum datasets called text similarity measure (TSM)

3 Proposed Method

Hamming distance is a metric for measure two binary data strings. While match two binary strings of the same length, Hamming distance is the number of bit positions in which the two bits are not the same. In the present work, hamming distance is applied to the data, which is not binary i.e. we have performed distance calculation based on the alphabet, present in the given word if both positions have a similar character the distance is zero else it is predicted to be one (Fig. 3).

As this word as an example:

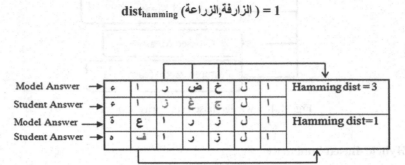

$$\text{dist}_{\text{hamming}} \ (\text{الزارفة,الزراعة}) = 1$$

Fig. 3. Hamming distance examples

The challenge in the hamming distance is it only measures the distance between same length strings. If the lengths of the two strings or comparing strings are different the distance calculation is worthless. So, the standard Hamming model considers that the answer is only correct when both the answers (student answer and model answer) have the same number of characters and there is no any difference (missing, wrong and added characters) in any position. Standard Hamming model also not considers the answer is correct if both the student answer and model answer have lexical similarity and they are derivatives of the same root such as يلعب(play), لعب(played), لاعب(player for male), لاعبة(player for female), ملعب(Playground), لاعبون(players), لعب(playing). Also, there is another issue related to the article of the word ال(the) in the Arabic language, the article ال(the) is always attached with the Arabic word such as اللعب(the playing), the standard Hamming model considers the answer is wrong for the student answer اللعب(the playing) that is not equal with model answer لعب(playing). The proposed Hamming model tried to solve these issues by applying the space padding pre-process to the answer that is smaller than the other answer. Padding space makes the text length of both answers is equal. Also, the proposed Hamming model tried to solve the similarity of unequal answer lengths by applying a space padding pre-process Also, the proposed Edit based model tried to solve issue of the lexical similarity for derivatives of the answer by applying another pre-process called stemming pre-process for both the student and model answers (Figs. 4 and 5).

3.1 Data Collection

To collect the required data for the proposed Hamming model, we designed some questions in Arabic language and stored in a *dictionary of questions* such as (ماهي الحرفة التي كان يعمل بها اجدادنا القدماء في اليمن ؟ What is the craft that our ancient ancestors used to work in Yemen?). We supposed the typical answer (model answer) for this question is الزراعة(farming), and we had a number of students answer the question (approximately 60 students), 20 students answered الزراعة(farming), 10 students

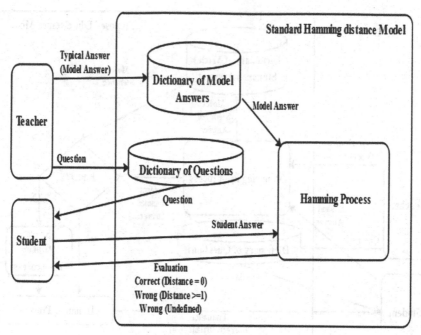

Fig. 4. Standard hamming distance

answered مزرعة(farm), 7 students answered مزارعون(farmers), 3 students answered مزارع(farmer), 10 students answered رياضة(sport) and 10 students answered ا رز(the word is missing letters), and so on for other questions.

3.2 Pre-processing Stage

The proposed model used two pre-processes: *stemming pre-process* and *space padding pre-*process. The stemming pre-process takes both the answers for the student and the model to remove the article ال(the) from the answer, and to return all derivatives of the word (the answer) to the root of the word. The derivatives of the answer have lexical similarity with the same meaning. The *stemmed model answer* will be stored in the *dictionary of the model answers*. The proposed model will use the root of the word to make exactly lexical similarity between the student answer and typical answer. The space padding is applied to the small length of the answer to become equal in length to the other answer. To understand the padding approach, we will investigate cases of standard Hamming distance model as follows:

There are three cases for the standard Hamming distance:

- *First case (normal case):* when the lengths of both texts are equal. For example, text1 has 7 characters, and text2 has also 7 characters. So, the Hamming distance will have 2 parameters as follows:

$$Hamming(text1, text2)$$

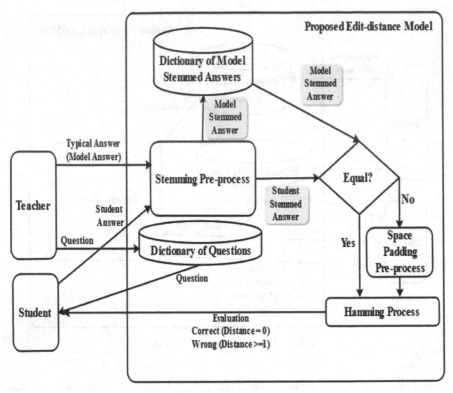

Fig. 5. Proposed edit distance.

Hamming operation will take the first parameter (text1) to compute its length (Here 7), and makes 7 comparisons between the characters of the two texts. The comparisons here are *possible* and *sufficient* to get *correct defined Hamming distance* value between 0 to 7 because the text1 string is equal in length to the text2 string. So, there is no need to a space padding where both the texts have the same length.

- *Second case (issue case):* when the lengths of both texts are un-equal. For example, text1 has 5 characters, and text2 has 7 characters. So, the Hamming operation will make 5 comparisons between the characters of the two texts. The comparisons here are *possible* but *insufficient* to get correct defined Hamming distance, where *wrong defined Hamming distance* value between 0 to 5 is produced, because the available comparisons (5 comparisons) are lesser than the text2 length (7 characters). So, there is need to two space paddings for text1 to be both the texts have the same length (7 characters).

- *Third case (second issue):* Also, when the lengths of both texts are un-equal. For example, text1 has 7 characters, and text2 has 5 characters. So, the Hamming operation will make 7 comparisons between the characters of the two texts. The comparisons here are *impossible*, so *undefined Hamming distance* is occurred due to a *string index out of range,* where the required comparisons (7 comparisons) are greater than the text2 length (5 characters). So, there is need to two space paddings for text2 to be both the texts have the same length (7 characters).

The space padding pre-process is not needed if both the answers have exactly the same root characters, but it is necessary if both the answers have different root characters.

3.3 Processing Stage

The main process of the proposed process is the ***Hamming process*** that takes both the answers of the student and the model as ***stemmed answers*** without space padding in case of the same root resulted from derivatives of the typical answer, and with space padding in case of different roots resulted from different answers to perform the Hamming operation for strings. Therefore, there is no possibility that the two answers are different in length; also, in most cases the proposed model returns 0 value as a distance for derivative words. The proposed model gives equal or greater than 1 as a distance if there is missing or/and wrong in characters.

3.4 Algorithm of the Proposed Hamming Model

Step 1: Give the question to the student from questions dictionary.
Step 2: Take the student and model answers and perform Stemming process   on them.
        Set st = ISRIStemmer()
        Perform stdent_answer = st.stem(input("Answer: "))
        Perform Stemming operation manually for model answer
        and stored in model answer dictionary.
Step 3: perform the space padding on the stemmed answers
        if len(stdent_answer)>len(value):
            m=len(stdent_answer)-len(value)
            value=value+(" "*m)
        if len(stdent_answer)<len(value):
            m=len(value)-len(stdent_answer)
            stdent_answer=stdent_answer+(" "*m)
Step 4: Perform Hamming operation on the stemmed padded answers
        r=hammingDist(stdent_answer,value)
Step 5: Hamming distance returns 0 value for correct answer and >=1 for wrong answer

---

4   Result and Discussion

The result of the proposed edit-based model that is applied on some questions with 60 students is shown in the table below. The standard Hamming model achieved 33.3% correct answer with 0 correct defined distance value, and achieved 58.3% wrong answer with greater than or equal to 1 wrong defined distance value, and achieved 8.3% wrong answer with undefined distance value. But The proposed edit-based model achieved 75% correct answer with 0 correct defined distance value, and achieved 25% wrong answer with greater than or equal to 1 correct defined distance value (Figs. 6 and 7).

Table 2. Comparison between standard Hamming method and proposed Hamming method.

Number of Students	Student answer	Distance		Evaluation	
		standard Hamming model	proposed Edit based model	standard Hamming model	proposed Edit based model
20	الزراعة (7 characters)	0	0	Correct	Correct
10	مزرعة (5 characters)	5	0	Wrong	Correct
7	مزارعون (7 characters)	6	0	Wrong	Correct
3	مزارع (5 characters)	4	0	Wrong	Correct
5	رياضة (5 characters)	5	3	Wrong	Wrong
10	زرا (3 characters)	3	1	Wrong	Wrong
5	المزروعات (8 characters)	Undefined	0	Wrong	Correct

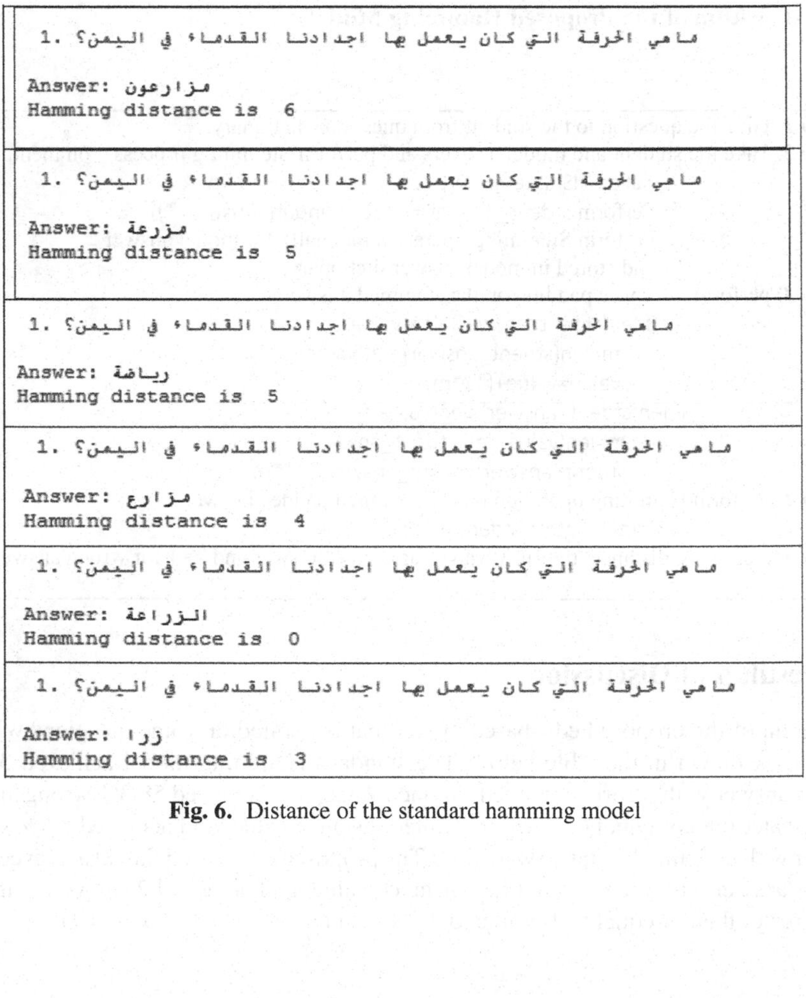

The table question row:
ماهي الحرفة التي كان يعمل بها اجدادنا القدماء في اليمن ؟ (question)
model answer: الزراعة (7 characters) (farm)

Fig. 6. Distance of the standard hamming model

Fig. 7. Distance of the proposed edit based model

5. Conclusion and Future Work

The result of the proposed edit-based model that is applied on some questions with 60 students it's in details on Table 2. The standard Hamming model achieved 33.3% correct answer with 0 correct defined distance value, and achieved 58.3% wrong answer with greater than or equal to 1 wrong defined distance value, and achieved 8.3% wrong answer with undefined distance value. But The proposed edit-based model achieved 75% correct answer with 0 correct defined distance value, and achieved 25% wrong answer with greater than or equal to 1 correct defined distance value.

References

1. Hammo, B., Abuleil, S., Lytinen, S., Evens, M.: Experimenting with a question answering system for the Arabic language. Comput. Humanit. **38**(4), 397–415 (2004). https://doi.org/10.1007/s10579-004-1917-3
2. Arai, K., Handayani, A.: Question answering system for an effective collaborative learning. Int. J. Adv. Comput. Sci. Appl. **3**, 60–64
3. Allam, A., Haggag, M.: The question answering systems: a survey. Int. J. Res. Rev. Inf. Sci. (IJRRIS) **2**(3) (2012)
4. Ray, S.K., Shaalan, K.: A review and future perspectives of Arabic question answering systems. IEEE Trans. Knowl. Data Eng. **28**, 3169–3190 (2016)

5. Bakari, W., Bellot, P., Neji, M.: Literature review of Arabic question-answering: modeling, generation, experimentation and performance analysis. In: Andreasen, T., et al. (eds.) Flexible Query Answering Systems 2015, vol. 400, pp. 321–334. Springer, Cham (2016). https://doi.org/10.1007/978-3-319-26154-6_25

6. Mishra, A., Jain, S.K.: A survey on question answering systems with classification. J. King Saud Univ. Comput. Inf. Sci. **28**, 345–361 (2016)

7. Gravano, L., et al.: Approximate string joins in a database (almost) for free. In: VLDB, vol. 1, pp. 491–500 (2001). http://www.vldb.org/conf/2001/P491.pdf

8. Yu, M., Li, G., Deng, D., Feng, J.: String similarity search and join: a survey. Front. Comp. Sci. **10**(3), 399–417 (2015). https://doi.org/10.1007/s11704-015-5900-5

9. Bhattacharya, A.: On a measure of divergence of two multinominal populations Sankhya. Indian J. Stat. **7**, 401–406 (1946)

10. Jaccard, P.: Étude comparative de la distribution florale dans une portion des Alpes et des Jura. Bull. Soc. Vaudoise. Sci. Nat. **37**, 547–579 (1901)

11. Krause, E.F.: Taxicab Geometry: An Adventure in Non-Euclidean Geometry. Courier Corporation (1975)

12. Friedman, J.H.: On bias, variance, 0/1—loss, and the curse-of-dimensionality. Data Min. Knowl. Discov. **1**(1), 55–77 (1997). https://doi.org/10.1023/A:1009778005914

13. Landauer, T.K., Dumais, S.T.: A solution to Plato's problem: the latent semantic analysis theory of acquisition, induction, and representation of knowledge. Psychol. Rev. **104**(2), 211–240 (1997). https://doi.org/10.1037/0033-295X.104.2.211

14. Gabrilovich, E., Markovitch, S.: Computing semantic relatedness using Wikipedia-based explicit semantic analysis. In: IJcAI, vol. 7, pp. 1606–1611 (2007). http://www.aaai.org/Papers/IJCAI/2007/IJCAI07-259.pdf

15. Mihalcea, R., Corley, C., Strapparava, C., et al.: Corpus-based and knowledge-based measures of text semantic similarity. In: AAAI, vol. 6, pp. 775–780 (2006). http://www.aaai.org/Papers/AAAI/2006/AAAI06-123.pdf

16. Budanitsky, A., Hirst, G.: Evaluating WordNet-based measures of lexical semantic relatedness. Comput. Linguist. **32**(1), 13–47 (2006). https://doi.org/10.1162/coli.2006.32.1.13

17. Monge, A.E., Elkan, C., et al.: The field matching problem: algorithms and applications. In: KDD, pp. 267–270 (1996). http://www.aaai.org/Papers/KDD/1996/KDD96-044.pdf

18. Wang, J., Li, G., Fe, J.: Fast-join: an efficient method for fuzzy token matching based string similarity join. In: 2011 IEEE 27th International Conference on Data Engineering, pp. 458–469 (2011). https://doi.org/10.1109/ICDE.2011.5767865

19. Lin, C., Liu, D., Pang, W., Wang, Z.: Sherlock: a semi-automatic framework for quiz generation using a hybrid semantic similarity measure. Cogn. Comput. **7**(6), 667–679 (2015). https://doi.org/10.1007/s12559-015-9347-7

20. Al-Hassan, M., Lu, H., Lu, J.: A semantic enhanced hybrid recommendation approach: a case study of e-Government tourism service recommendation system. Decis. Support Syst. **72**, 97–109 (2015). https://doi.org/10.1016/j.dss.2015.02.001

21. Atoum, I., Otoom, A.: Efficient hybrid semantic text similarity using WordNet and a corpus. Int. J. Adv. Comput. Sci. Appl. **7**(9), 124–130 (2016). https://doi.org/10.14569/IJACSA.2016.070917. http://thesai.org/Publications/ViewPaper?Volume=7&Issue=9&Code=ijacsa&SerialNo=17

Ontology Model for Spatio-Temporal Contexts in Smart Home Environments

L. Shrinidhi⑩, Nalinadevi Kadiresan(✉)⑩, and Latha Parameswaran⑩

Department of Computer Science and Engineering, Amrita School of Engineering, Amrita Vishwa Vidyapeetham, Coimbatore, India
{k_nalinadevi,p_latha}@cb.amrita.edu

Abstract. Smart home environment supports in simplifying the daily routines of the residents by learning the repetitive tasks and automating the activities. Sensors provide an unobtrusive way of collecting the state change in the environment, residents and the objects. The numbers of sensors are directly proportional to the cost and power consumptions. The sensor to activity mapping can be used for various task in smart home environment like the sensor optimization and sensor placement. Though the data-driven methods are proven to provide accurate results for recognizing activities, it does not provide context information for sensor to activity mapping. This paper deals with identifying sensors used in an activity, based on the spatial and temporal contexts. An ontology model is developed for representing the real-time smart home sensor data. A rule-based reasoner is developed using SWRL and SQWRL to infer spatial and temporal contexts. In SWRL rules, spatial context provides insight on where an activity happens. This becomes vital when more than one activity takes place at two different places. Thereby, the sensors responsible for monitoring an activity during the occurrence of concurrent events are derived. Similarly, with the help of temporal information, the path covered by the user when performing an activity is traced. The results from the developed expert system serve as input for sensor optimization task.

Keywords: Ontology model · Sensor optimization · Spatio-temporal context · Smart home

1 Introduction

With the advancement in engineering and technology, ambient intelligence has become an interesting topic over a decade [1] in smart environments. Smart home environments make the day-to-day surrounding sensitive to people through sensors [2, 3]. The activities of the residents can be monitored and captured using ambient sensors, visual sensors, wearable sensors and smartphones [4]. Monitoring user's actions provides information on the resident's whereabouts and the actions that are carried out at a particular period of time. Such a kind of information becomes essential for applications like, remote health monitoring, support for a senior citizen who lives independently, user's behavior detection, etc.

© IFIP International Federation for Information Processing 2021
Published by Springer Nature Switzerland AG 2021
V. Krishnamurthy et al. (Eds.): ICCIDS 2021, IFIP AICT 611, pp. 113–124, 2021.
https://doi.org/10.1007/978-3-030-92600-7_11

The data collected from sensors provides minimal knowledge as the incoming data is a set of raw values. In a multi resident smart home environment, the daily activities captured by the sensors has high volume and velocity. Each activity however is a collection of actions indicated by raw sensor values. Thus, to represent sensor data in a more meaningful manner, this work focuses on developing an Ontology based knowledge representation. The relationship between sensor data and the activities are represented as properties in the ontology model and inferred using SWRL (Semantic Web Rule Language) and SQWRL (Semantic Query-enhanced Web Rule Language) rules. This paper focuses on finding suitable inferences using rule-based reasoning with help of spatial context. The results of inference mechanism will serve as an input for various applications like, sensor optimization, activity recognition, user behavior analysis, sensor selection for efficient power consumption, etc.

The idea of inferencing based on spatial context is to provide additional information to context-aware applications. For example, to recognize an activity, information about the time and active sensors alone are not sufficient to accomplish the task. The whereabouts of the activity also plays a major role. Similarly, in a multi-resident living, during the occurrence of concurrent activities it becomes tedious to find the sensors responsible for monitoring the activity. This work uses spatial context to derive set of sensors that are responsible for recognizing an activity in a multi-resident smart home. Additionally, temporal context is used to trace the path covered by the user while performing an activity.

This paper is organized as follows: Sect. 2 provides detailed literature survey of recent advancement in this area of study; Sect. 3 depicts the working principle of process carried out and the dataset used for this study. Section 4 depicts the dataset used. Section 5 explains the concepts and technologies used for knowledge-based construction and rule-based reasoning. Section 6 provides the experimental results of rules used for inferencing. The proposed method is concluded in Sect. 7.

2 Literature Survey

Many researchers have explored and developed ontology models to represent context knowledge for a smart home environment. A few such prominent contributions have been discussed here.

To describe real-world data and to build context-aware applications, the ontology model should express hierarchical structure of sensors, user's behavioral information and observational results. Since ontologies help to discover the hidden relationship that are left unrecognized, the relationship between the concepts provides a major contribution to knowledge representation [5]. Numerous ontology-based knowledge representations are used to describe these concepts and relationships. The languages include Resource Description Framework (RDF), Web Ontology Language (OWL), Ontology Inference Layer (OIL) which allow the semantics understandable to machines.

Surveys are conducted to determine the need and fundamentals of ontologies in the area of context-aware applications [6, 7]. In [8], a Smart Home Ontology (SHO) was developed for home automation system, SHO ontology is built on Semantic Sensor Network (SSN) thereby promoting reusability. The main aim of constructing SHO for a home automation system is to cover all the functionalities with minimum number of classes and properties. One main drawback of using SHO is, it does not provide the user's state at a time.

The study mentioned in [9] devices an ontology model for deploying software applications on suitable devices with help of context information. To accomplish the task of deploying suitable software applications, it considers user capabilities, device configurations, software needs, and location preferences. In [10], a healthcare ontology using wearable sensors was designed focusing on customized healthcare services. With the help of SSN ontology, residential building monitoring [11] was done to monitor the unhealthy exposure of carbon monoxide. In [12], more contextual information is added to the developed ontology but in the prospects of visual information gathered from visual sensors. These ontologies were domain-specific and application-specific. It also fails to provide knowledge sharing and reusability.

The proposed work focuses on developing an ontology that connects all the aspects of observational values (like sensor data, value, time of occurrence, etc.,) along with the user responsible for it. Since the proposed work is based on concepts facilitated by the SOSA (Sensor, Observation, Sample and Actuator) ontology model, it also promotes the idea of reusability. Additional properties are also included to improve the semantic relationship between concepts thereby adhering to the functionality of interoperability.

For context-aware applications, the existence of spatial and temporal context plays a vital role [13–15]. In [16] the authors explain the importance of spatial and temporal concepts in context-aware reasoning. It describes how machine learning and symbolic approaches are used to provide contextual reasoning from the given sensor data. The authors in [17] explain how qualitative reasoning is carried out using spatial and temporal data to monitor elderly activities in a smart home. The research in [18], uses spatial and temporal concepts for recognizing the habit of daily living in a smart home. It uses SWRL rules to generate simple and complex habits exhibited by the user. In [19] a detailed qualitative approach of spatiotemporal reasoning is depicted. This research mentions the use of Allen's temporal relation for tracing the path covered by the resident to accomplish a task. The mentioned work gives only a qualitative study of spatial and temporal reasoning and does not provide suitable insights for multi-resident living. The proposed work focuses on quantitative study of Allen's temporal relations on real-time smart home data. It also depicts how spatial context is used for distinguishing sensors during concurrent activities taking place in a multi-resident smart home.

3 Proposed Architecture

Figure 1 shows the architecture in building an expert system; it consists majorly of three modules: pre-processing of smart home data, construction of knowledge base and rule-based reasoning.

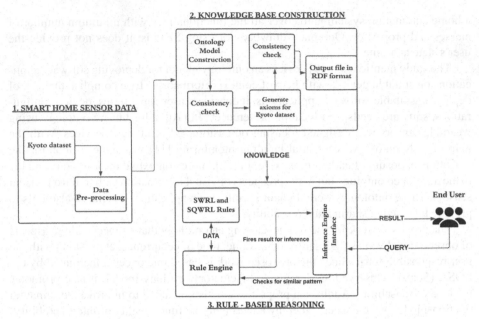

Fig. 1. Architecture diagram

3.1 Working Principle

The dataset used in this proposed work, follows Western State University (WSU) housing system [20]. Data pre-processing is applied to real-world smart home data. An Ontology model (using OWL) is constructed which acts as a semantic interoperable layer for representing the sensor data. This is accomplished by converting real-time entities into classes and their characteristics into properties. It is used to create common vocabularies and axioms in a machine-readable format.

After the successful creation of the ontology model, the consistency of the model is checked using reasoners. The reasoner helps to infer as set of logical relation from a set of facts. It also checks the compatibility of the axioms generated with the annotations defined. These set of facts represent the semantics about the smart home data. The facts and axioms are stored in Resource Description Framework (RDF) format and act as a Knowledge base for inferencing.

With the knowledge created, SWRL and SQWRL rules are developed to identify the sensors during the occurrence of concurrent events. To do so, the rules make use of spatial context; temporal context is also used for order of activity carried out by a user and for tracing a user's activity.

4   Dataset Description

A smart home is precise in its functionality and conforms that monitoring any activities does not affect the lifestyle of residents. Many residents might feel using vision-based sensors would invade their privacy. So, this research focuses on non-visual sensors monitoring the activities of the residents.

The dataset used for this research comes from a real-world smart environment deployed by the CASAS (Centre of Advanced Studies in Adaptive Systems) facility. The Kyoto dataset represent sensor values in a three-bedroom apartment and based on WSU (Washington State University) Housing System. Many machine learning algorithms [21] have been used on the CASAS dataset and proved to be effective for monitoring Activities of Daily Living (ADL) tasks.

The residents in the apartment are two undergraduate students performing their daily activities. In the smart home environment, sensors are deployed in living room, kitchen, closets, bedroom and bathroom. The sensors that are deployed in this smart apartment are motion sensors (M0##), light sensors (L0##), temperature sensors (T0##), door sensors (D0##), certain object sensors are also deployed on particular objects (I0##) and electrical usage (P001).

Kyoto dataset consists of 13 activities carried out by two residents (R1 and R2) in the year 2009–2010 where each sample in the dataset consists of the date and time of each event, sensor ID, and sensor status (numeric and binary) activated during an event. The dataset records concurrent activities happening in the household.

5 Construction of Expert System

5.1 Development of Ontology Model

Out of many sensor ontologies developed, W3C Semantic Incubator Group developed an OWL 2 ontology called as Semantic Sensor Network (SSN) Ontology that describes the capabilities of sensors, observed property, a feature of interest that is to monitor, and actuator properties, and much more. Michael Compton et.al [22] explains about Semantic Sensor Network (SSN) Ontology along with concepts and relationship established between the concepts.

SSN follows on two types of architecture namely, horizontal and vertical modularization by including lightweight Sensor, Observation, Sample, Actuator (SOSA) ontology as self-contained core ontology. The ontology model is built based on SOSA architecture thereby promoting reusability of existing concepts. The ontology model was developed based on the incoming sensor data from the Kyoto testbed and according to guidelines specified by Thomas R. Gruber [23]: explicit specification of concepts, usage of ontological commitments to communicate knowledge of an agent involved, consistency between concepts and relationship, and ability to share vocabulary.

Figure 2 demonstrates the set of classes and properties used for representing sensor data from smart environment.

The classes described in Fig. 2 exhibit similar vocabularies of Sensor, Observation, Sample, and Actuator (SOSA) Ontology model [24]. The class sosa:Resident is equivalent to sosa:FeatureOfInterest class of SOSA model.

The data property, owl:hasObservationValue represents the sensor value made for sosa:Observation, whereas owl:hasAction indicates the type of activity, and owl:hasOccurredAt denotes the timestamp of when sosa:Observation occurred. Other object properties have similar definitions of SOSA Ontology Model.

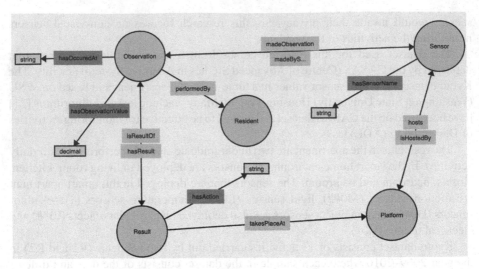

Fig. 2. Ontology model for representing sensor data

5.2 Importing Sensor Data as Individuals

With classes and relationships specified, the ontology model is created with help of Protégé tool. The developed ontology model is checked for consistency with help of HermiT reasoner. Only when a model is consistent, any logical relationship derived with instances of the ontology model will be stated to be true or valid.

Protégé comes within an inbuilt plugin called as Cellfie. The functionality of the Cellfie plugin is to convert instances in Kyoto dataset from an excel sheet into individuals of the ontology model. After loading the instances, to convert the samples into axioms of classes and properties, transformation rules are used. The role of transformation rules is to convert raw sensory data into individuals (i.e.,) converts the instances into Description Logic.

The mappings of data samples from excel sheet into OWL ontologies are done by MappingMaster which uses Domain Specific Language (DSL). With the help of DSL, any clause that is represented in Manchester Syntax indicating OWL class, OWL property, OWL individual can be replaced with a reference clause. The references in MappingMaster DSL are prefixed by '@' symbol, which is generally followed by an Excel-style cell.

> An example of transformation rule:
> Individual: @A*
> Types: Observation
> Facts: madeBySensor @C*, isPerformedBy @F*, hasResult @G*,
> hasOccuredAt @B*(xsd:dateTime), hasObservationValue @D*

Here, "Individual" represents the instances that are in the dataset. '@A*' denotes an Excel sheet reference cell and '*' specifies all the instances in the specified column. The above transformation rule states that all instances from column-A are of type Observation class. The corresponding properties associated with sosa:Observation class are defined in the Facts section. After loading the Transformation rules, a set of individuals and corresponding axioms are generated according to rules specified.

Figure 3 depicts the set of axioms generated successfully for the individuals mentioned in the transformation rules.

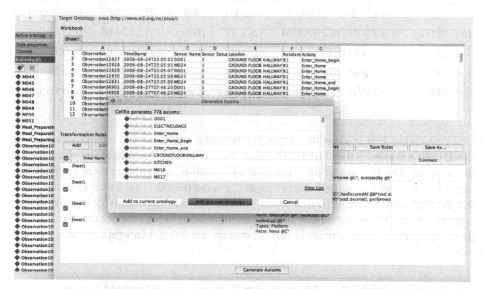

Fig. 3. Axioms generated for Individuals using Cellfie plugin

After successful creation of axioms, the ontology is stored in RDF format because the data becomes semantic which is excellent for data integration and interoperability. With the Knowledge Base constructed, rules are used to derive the desired results.

6 Results and Discussion

In the following sections, the notations described in Sect. 5 are used for creating SWRL and SQWRL rules.

6.1 Rule for Activity Based Sensor Identification – Without Spatial Context

sosa:Observation(?obs) ^ sosa:madeBySensor(?obs, ?sensor) ^ sosa:isHostedBy(?sensor, ?platform) ^ sosa:hasResult(?obs, sosa:Eating) ^ sosa:performedBy(?obs, ?res) ^ sosa:hasOccuredAt(?obs, ?time) -> sqwrl:selectDistinct(?sensor, ?platform, ?res, ?time, sosa:Eating) ^ sqwrl:orderBy(?time)

This rule retrieves all sensors that are activated when a resident is eating (i.e.,) sensors that are activated in the specified time window. The query produces the following result,

Table 1. Results produced for sensor identification – without spatial context

Sensor	Platform	Res	Time	c4
sosa:M005	sosa:LIVINGROOM	sosa:R1	2009-08-30T23:24:03	sosa:Eating
sosa:M034	sosa:RIGHTBEDROOM	sosa:R1	2009-08-30T23:24:04	sosa:Eating
sosa:M034	sosa:RIGHTBEDROOM	sosa:R1	2009-08-30T23:24:05	sosa:Eating
sosa:M034	sosa:RIGHTBEDROOM	sosa:R1	2009-08-30T23:24:06	sosa:Eating
sosa:M005	sosa:LIVINGROOM	sosa:R1	2009-08-30T23:24:07	sosa:Eating

From Table 1, it is clear that when R1 is eating some sensors are activated in the LIVINGROOM and RIGHTBEDROOM at the same time, this makes the above result inconsistent. The inconsistency is because, in a multi-resident living there are possibilities of more than one activity occurring at the same period. The sensor readings from RIGHTBEDROOM are activated when another person (say R2) is doing a different activity when R1 is eating. Thus, for consistent results, to find the sensors that are utilized when a particular activity is carried out, spatial context information plays a vital role.

6.2 Rule for Activity Based Sensor Identification – with Spatial Context

Step 1: sosa:Observation(?obs) ^ sosa:madeBySensor(?obs, ?sensor) ^ sosa:isHostedBy(?sensor, ?platform) ^ sosa:hasResult(?obs, sosa:Eating_begin) -> sosa:takesPlaceAt(sosa:Eating, ?platform)

The rule specified in Step 1 retrieves the spatial context of where an activity begins and stores it to the existing ontology. So even when a user performs the same activity in different locations, the SWRL rules fetches the information from a set of facts and assigns the spatial context information to ontology's object property "takesPlaceAt".

Step 2: sosa:Result(sosa:Eating) ^ sosa:takesPlaceAt(sosa:Eating,?platform) ^ sosa:isResultOf(sosa:Eating,?obs) ^ sosa:madeBySensor(?obs,?sensor) ^ sosa:isHostedBy(?sensor,?platform) ^ sosa:performedBy(?obs,?res) ^ sosa:hasOccuredAt(?obs,?time) -> sqwrl:select(?sensor,?platform,?time, ?res, sosa:Eating) ^ sqwrl:orderBy(?time)

It becomes a laborious process to distinguish between the sensors responsible for monitoring a set of activities in a multi-resident living, especially when more than one user performs different activities at the same timestamp. However, this rule associates a sensor reading to a specific user performing a specific activity. Table 2 depicts the result of successful execution of rules specified in Step 1 and 2, fetching only the sensor activated when the resident is eating.

Table 2. Results produced on execution of Step 1 and Step 2

Sensor	Platform	Time	Res	c4
sosa:M006	sosa:LIVINGROOM	2009-08-31T00:14:44	sosa:R1	sosa:Eating
sosa:M007	sosa:LIVINGROOM	2009-08-31T00:14:47	sosa:R1	sosa:Eating
sosa:M006	sosa:LIVINGROOM	2009-08-31T00:14:48	sosa:R1	sosa:Eating
sosa:M007	sosa:LIVINGROOM	2009-08-31T00:14:49	sosa:R1	sosa:Eating
sosa:M007	sosa:LIVINGROOM	2009-09-01T07:27:45	sosa:R2	sosa:Eating
sosa:M009	sosa:LIVINGROOM	2009-09-01T07:27:46	sosa:R2	sosa:Eating
sosa:M015	sosa:LIVINGROOM	2009-09-01T07:27:47	sosa:R2	sosa:Eating

When temporal value is sorted in ascending order, the path covered by user while performing an activity is traced. On finding the sensors utilized, these results have the potential to act as input for sensor optimization applications, activity recognition tasks, and for sensor selection scenarios to minimize power consumption.

6.3 Rule for Identifying Order of Activity: With Temporal Context

With the developed ontology model and with help of SWRL rules, user's behaviour can also be inferred using spatial and temporal context. On making use of Allen's temporal relations [25] with SWRL rules, different combination of results can be inferred. An example of deriving the amount of time taken by the user to wander after sleep is obtained using the following SWRL rule with Allen's temporal relation temporal:after(),

sosa:Observation(?obs) ^ sosa:hasOccuredAt(?obs, ?sleep_end) ^ so-
sa:hasResult(?obs, sosa:Sleep_end) ^ so-
sa:takesPlaceAt(Sleeping,?platform)^sosa:performedBy(?obs,R1)
^sosa:Observation(?obs1)^sosa:hasOccuredAt(?obs1,?wander_begin)^sosa:hasResult
(?obs1,sosa:Wandering_in_room_begin)^sosa:takesPlaceAt(Wandering_in_room,
?platform1)^sosa:performedBy(?obs1,R1)^temporal:after(?wander_begin,?sleep_end)
^temporal:duration(?d,?sleep_end,?wander_begin,"Minutes")->
sqwrl:select(?sleep_end, ?wander_begin, ?d) ^ sqwrl:orderBy(?sleep_end)

The results of successful execution of the rule created for finding the order of activity are depicted in Table 3. The time difference between two events is described in minutes.

Qualitative analysis of spatiotemporal reasoning in the literature does not provide suitable insights on how the rule behaves in a multi-resident living space. To overcome

Table 3. Results produced for identifying order of activity

sleep_end	wander_begin	d
2009-08-24T00:05:21	2009-08-24T00:16:29	"11"xsd:long
2009-08-24T00:10:14	2009-08-24T00:16:29	"6"^^xsd:long
2009-08-24T00:16:27	2009-08-24T00:16:29 .	"6"^^xsd:long
2009-08-26T06:18:22	2009-08-26T07:07:09	"49"^^xsd:long
2009-08-28T06:29:13	2009-08-28T07:01:00	"32"^^xsd:long

those constraints the SWRL and SQWRL rules are created based on spatial and temporal context on real-time smart environment and provide the following inferences that also gives suitable insights for many Ambient Intelligence applications,

- With spatial context, differentiating concurrent activity occurrence by different users in same timestamp is established.
- Using spatial and temporal context, the pattern of user performing an activity and the order of the activity (i.e., a MEAL PREPARATION is generally followed by an EATING activity) can be interpreted (using Allen's temporal relations).

With statistical analysis, the mapping of sensors activation when an activity being performed is probabilistic due to lack of contextual information. To enhance the activity-sensor mapping, the spatial and temporal information are embedded in the ontology model. This contextual information helps in identification of sensors being triggered for every action undergone by the user to complete an activity. With spatial and temporal context used in the ontology model, the results say that, sensor S is activated when an activity A is performed by the resident R for every timestamp between begin and end of an activity. Thus, the results produced eventually leads to higher precision and recall scores.

7 Conclusion and Future Work

In this article, we have discussed an ontology-based knowledge representation for smart home sensor data that was built for performing logical inferences with the help of spatio-temporal context. Smart home sensor data are semantically annotated with the help of SOSA ontology. The instances from Kyoto's smart environment are converted into individuals and axioms of the ontology. The rule-based reasoner allows to perform logical inferences from existing facts. SWRL and SQWRL rules were developed to infer logical relationships that exist but remained unrecognized with help of spatial and temporal contexts from ontology. The developed rules are validated with the data derived from the Kyoto smart home environment. The rules make use of spatial and temporal data from smart home environment to provide consistent results on sensor identification, order of activity identification and tracking user's activities. As ontology models are reusable and extendible, new domain concepts can also be added to enrich the knowledge base of an expert system. The spatial and temporal contexts obtained from the reasoner could be further used in sensor optimization task.

References

1. Ramos, C., Augusto, J.C., Shapiro, D.: Ambient intelligence -the next step for artificial intelligence. IEEE Intell. Syst. **23**(2), 15–18 (2015)
2. Nagarajan, B., Shanmugam, V., Ananthanarayanan, V., Bagavathi Sivakumar, P.: Localization and indoor navigation for visually impaired using bluetooth low energy. In: Somani, A.K., Shekhawat, R.S., Mundra, A., Srivastava, S., Verma, V.K. (eds.) Smart Systems and IoT: Innovations in Computing. SIST, vol. 141, pp. 249–259. Springer, Singapore (2020). https://doi.org/10.1007/978-981-13-8406-6_25
3. Dharan, B., Kumar K., Akshaya Srinivasan, R., Vidhya, S.: Smart home based interactive indoor navigation system. J. Adv. Res. Dyn. Contr. Syst., 953–956 (2018)
4. Nandy, J.S., Chowdhury, C., Singh, K.P.D.: Detailed human activity recognition using wearable sensor and smart phones. In: 2019 International Conference on Opto-Electronics And Applied Optics (OPTRONIX), Kolkata, India, pp. 1–6 (2019)
5. Gowtham, R., Venugopal, A.: A study on verbalization of OWL axioms using controlled natural language. Int. J. Appl. Eng. Res. **10**(7), 16953–16960 (2015)
6. Baldauf, M., Dustdar, S., Rosenberg, F.: A survey on context aware systems. Int. J. Ad Hoc Ubiqit. Comput. **2**(4), 263–277 (2007)
7. Perera, C., Zaslavsky, A., Christen, P., Georgakopoulos, D.: Context aware computing for the internet of things: a survey. IEEE Commun. Surv. Tutor. **16**(1), 414–454 (2014)
8. Berat Sezer, O., Can, S.Z., Dogdu, E.: Development of a smart home ontology and the implementation of a semantic sensor network simulator: an Internet of Things approach. In: 2015 International Conference on Collaboration Technologies and Systems (CTS), Atlanta, GA, pp. 12–18 (2015)
9. Vallerand, C.G., Abdulrazak, B., Giroux, S., Anind, D.: A context-aware service provision system for smart environments based on the user interaction modalities. J. Ambient Intell. Smart Environ. **5**(1), 47–64 (2013)
10. Kim, J., Kim, J., Lee, D., et al.: Ontology driven interactive healthcare with wearable sensors. Multimed Tools Appl. **71**, 827–841 (2014)
11. Stocker, M., Rönkkö, M., Kolehmainen, M.: Making sense of sensor data using ontology: a discussion for residential building monitoring. In: Iliadis, L., Maglogiannis, I., Papadopoulos, H., Karatzas, K., Sioutas, S. (eds.) AIAI 2012. IAICT, vol. 382, pp. 341–350. Springer, Heidelberg (2012). https://doi.org/10.1007/978-3-642-33412-2_35
12. Daoutis, M., Coradeschi, S., Loutfi, A.: Grounding common-sense knowledge in intelligent systems. Ambient Intell. Smart Environ. **1**(4), 311–321 (2009)
13. Menon, V., Jayaraman, B., Govindaraju, V.: Enhancing biometric recognition with spatio-temporal reasoning in smart environments. J. Pers. Ubiquit. Comput. **17**, 987–998 (2013)
14. Menon, V., Jayaraman, B., Govindaraju, V.: Probabilistic spatio-temporal retrieval in smart spaces. J. Ambient Intell. Hum. Comput. **5**(3), 383–392 (2014)
15. Menon, V., Jayaraman, B., Govindaraju, V.: Spatio-temporal querying in smart spaces. In: Proceedings of 3rd International Conference on Ambient Systems, Networks and Technologies (ANT-2012), Ontario, Canada, vol. 10, pp. 366–373 (2012)
16. Guesgen, H.W., Marsland, S.: Spatio-temporal reasoning and context awareness. In: Nakashima, H., Aghajan, H., Augusto, J.C. (eds.) Handbook of Ambient Intelligence and Smart Environments, vol. 4, pp. 609–634. Springer, Boston (2010). https://doi.org/10.1007/978-0-387-93808-0_23

17. Sioutis, M., Alirezaie, M., Renoux, J., Loutfi, A.: Towards a synergy of qualitative spatio-temporal reasoning and smart environments for assisting the elderly at home. In: 30th International Workshop on Qualitative Reasoning (Held in Conjunction With IJCAI 2017), Melbourne, Australia (2017)

18. Wang, P., Luo, H., Sun, Y.: A habit-based SWRL generation and reasoning approach in smart home. In: 2015 IEEE 21st International Conference on Parallel and Distributed Systems (ICPADS), Melbourne, pp. 770–775 (2015)

19. Gottfried, B., Guesgen, H., Hübner, S.: Spatiotemporal reasoning for smart homes. In: Augusto, J Carlos, Nugent, C.D. (eds.) Designing Smart Homes. LNCS (LNAI), vol. 4008, pp. 16–34. Springer, Heidelberg (2006). https://doi.org/10.1007/11788485_2

20. Centre of Advanced Studies in Adaptive Systems (CASAS). http://casas.wsu.edu/datasets/. Accessed 14 Nov 2020

21. De-La-Hoz-Franco, E., Ariza-Colpas, P., Quero, J.M., Espinilla, M.: Sensor-based datasets for human activity recognition – a systematic review of literature. IEEE Access **6**, 59192–59210 (2018)

22. Compton, M., et al.: The SSN ontology of the W3C semantic sensor network incubator group. J. Web Semant. **17**, 25–32 (2012)

23. Gruber, T.R.: Toward principles for the design of ontologies used for knowledge sharing. Int. J. Hum.-Comput. Stud. **43**(5–6), 907–928 (1995)

24. Semantic Sensor Network Ontology W3C Recommendation. https://www.w3.org/TR/vocab-ssn/

25. Batsakis, S., Stravoskoufos, K., Petrakis, E.: Temporal reasoning for supporting temporal queries in OWL 2.0. In: König, A., Dengel, A., Hinkelmann, K., Kise, K., Howlett, R.J., Jain, L.C. (eds.) KES 2011. LNCS (LNAI), vol. 6881, pp. 558–567. Springer, Heidelberg (2011). https://doi.org/10.1007/978-3-642-23851-2_57

Automatic Detection of Buildings Using High Resolution Images for Medium Density Regions

Karuna Sitaram Kirwale$^{(\boxtimes)}$, Seema S. Kawathekar, and Ratnadeep R. Deshmukh

Department of Computer Science and IT, Dr. Babasaheb Ambedkar Marathwada University,
Aurangabad, India
rrdeshmukh.csit@bamu.ac.in

Abstract. As the construction material of the building is different hence building detection is critical task from the HRI. The image qualities, resolution, weather type of image sensor are important factors to produce accuracy. Building detection in dense urban areas having problems due to factors like shape, size, color and texture, and image sensor. For the find, the building has to consider the characters of the building like contrast, shape, and the building allocation in high, low, and medium density. In the present study, the mathematical morphological operation is used for the separation of the building. The building is indicated with a boundary.

Keywords: Automatic building detection · Morphological operation

1 Introduction

Automatic extraction of the building using remotely sensed data give limitations on the resulting performance. When the high-resolution image was studied for the building exaction it shows the different complexity in a scene that is due to the low and poor contrast or the same spectral reflectance of the objects that appeared in the image. This gives difficulties for the identification of the building. Again when building roofs are considered building structured to be found that they are varying spectral properties with building construction are diverse, slope angles and flatted building roofs. If the image appears with the verity in construction of buildings then the given flow of algorithm may have problems in exact identification of the building. Self-occlusion in building rooftop and the shadow of building proposed challenges for extraction. Another limitation comes from the same spectral reflection of the concrete building roof and concrete road. The same things happen with concrete parking area available in the urban areas. A sometime non-building objects which is man-made were considered as building and which gives false accuracy. By studying high-resolution images with different modality as dense, medium, and high density the picture gives the information that if the buildings are very closely constructed then that buildings are considered as a single building where it found many buildings are available there. For example row houses, colonies with very close construction, apartments, etc.

Automatic extraction of the building applicable for urban planning, disaster management, flood assessment, taxes fixing, urban development planners, military, etc.

© IFIP International Federation for Information Processing 2021
Published by Springer Nature Switzerland AG 2021
V. Krishnamurthy et al. (Eds.): ICCIDS 2021, IFIP AICT 611, pp. 125–131, 2021.
https://doi.org/10.1007/978-3-030-92600-7_12

2 Literature Survey

Building extraction techniques using high-resolution satellite images can classify with the help of automatic and semi-automatic techniques.

Working with aerial images to find rectangular buildings with flat roofs with geometric shapes, projection constraints from single-intensity images use edge detection-based technology [1]. Various technologies have been compared and evaluated based on restrictive standards such as evaluation and quality, information, and quality. These technologies combined for extraction of object and development of aerial image for object extraction basic problems [2]. For large building extraction and escaping shadows of buildings from high-resolution QUICKBIRD panchromatic images edge detection based techniques are applied. This method did not extract petite buildings with little or no shadows. To extract small buildings [3], the spectral information with the development of structural and discourse details and used the image of the IKONOS satellite in Columbia, Missouri. Within the technical scope of the plan, structure, and discourse details are habitually used to differentiate between buildings and parking spaces and alternatives with similar spectral information [4]. Advanced morphological operators, such as Hit or Miss Transforms with variable sizes and structured forms, extract buildings from High-Resolution QUICKBIRD panchromatic images. The accuracy of this method was calculated based on the accuracy rate, which is half of 1 mile, and the letter value of the letter is 63. To improve the overall accuracy found for building extraction used the "morphological shading index" (MSI) [5], with the "morphological building index" (MBI) that gives building count [6] in an object-based framework [7]. Besides, retrieved the imaginative buildings of the HRS GeoEye-1 shopping mall in Washington, D.C., using the different shape shadow and building operator [8]. The overall accuracy (OA) of the planning method is 95.12%. Multi-index learning (MIL) [9] techniques for HRS picture development index sets, such as MBI, MSI, and Normalized Vegetation Index (NDVI) are used to enhance the classification results of urban areas. To improve classification accuracy, develop remote sensing images. The Generalized Differential Morphological Profile (GDMP) [10] and found that it is superior to the ancient Differential Morphological Profile (DMP) [11]. Preprocessing DMP is used for classification by the neural network, the processing load is reduced by using neural network call boundary feature extraction, discriminant analysis feature extraction, and easy classification feature selection [12]. Their approach is supported by the idea that the building has a rectangular form, which is incorrect in fashionable urban affairs [13]. Identifying the rectangular and circular buildings from panchromatic high-resolution images and pan-sharpened IKONOS images the Hough transform and support vector machine (SVM) classification techniques are used [14]. To improve the High-Resolution Image classification accuracy, the supportive methods like C voting, P fusion, and OBSA are mixed with the spectrum, structure, and linguistic options for support vector machine (SVM) [15]. Also, the fully sharp image developed with building extraction accuracy is 5 shots larger than the MS image. However, despite the considerable space at the bottom of these buildings, the system cannot extract buildings with little or no shadows [16]. Object-based [17–20] methods. Building inspection formula, and plan a comprehensive analysis strategy for building inspection.

3   Data

Google Earth Pro images of an urban area in Aurangabad. These images cover the area named Sangita Colony, Samarth Nager, N8, Sanjivani Nager, Nath Nager, Vasundhara colony, Vedant Nagar, and Khadkeshwar which are situated in different location of the city. These images found buildings of different sizes, roof covering, forms, and arrangements. The high-resolution image with 3 bands Red (R), Green (G), and Blue (B).

4   Building Extraction

The process of building exaction can divide in to three Sect. 1 Preprocessing, Sect. 2. separating buildings from the background and Sect. 3 will count the building and building area (Fig. 1). The proposed method of flow is given below:

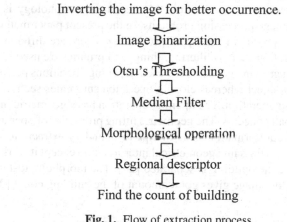

Inverting the image for better occurrence.

Image Binarization

Otsu's Thresholding

Median Filter

Morphological operation

Regional descriptor

Find the count of building

Fig. 1. Flow of extraction process

5   Preprocessing

In preprocessing the image converted to binary image and then after the image is complement. And the resulted image use for further process.

6   Otsu's Thresholding

Thresholding is an image segmentation technique basic idea of thresholding is to select an optimal gray-level threshold value for separating objects of interest in an image from the background based on their gray-level distribution. While manually easily differentiable an object from the complex background and image thresholding is a difficult task to separate them. To develop thresholding algorithms of an image, the gray-level histogram of an image is used efficiently as a tool [21, 22, 23, 24].

Otsu's Thresholding method is simple and effective, which comes from global thresholding and that only represents gray value of the image. The Otsu method was proposed by Scholar Otsu in 1979 [25]. Two dimensional Otsu algorithms were works on both the gray-level threshold of each pixel and its spatial correlation information between neighborhood. So for noisy images Otsu algorithm gives satisfactory segmentation results [26]. From few decade the thresholding is effective technique for better results. OTSU'S Method works based on the very simple idea that minimizes the weighted within-class variance. This turns out to be represented the same as maximizing the between-class variance. Once the histogram is constructed for the given image, it works directly on the gray level histogram [27].

7 Morphological Operations

The mathematical morphology is a tool that helps to represent and describe region shapes, like boundaries, skeleton, and convex hull. Mathematical morphology is a powerful approach to multiple image processing problems. In the present paper mathematic morphology is used for preprocessing and post-processing. There are different techniques of morphological operation such as filtering, thing, and pruning are used [24].

For that, the region growing concepts are used. Opening operations perform smoothing the contours of an object whereas closing operation smoothies section of contours but opposite to opening operation. The opening operation has a geometric interpretation that performs rolling ball concepts. The geometric fitting properties of opening operation gives to a set of theoretic formation which helps to boundary extraction of an object. The closing operation has the same geometrical interpretation except it works on outside boundaries. And hence the building is separated [24]. The morphological operation on 30 cm GeoEye-I satellite image gives good amount of the building count [28].

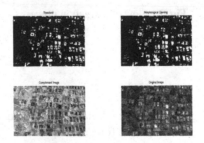

Fig. 2. Samarth nager

8 Results

Figure 2 and Fig. 3 show the result with the output as building both the images show the medium density images.

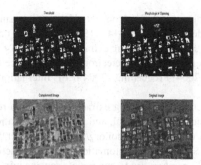

Fig. 3. Sangita colony

The first step to convert the image to the complemented image and find the threshold values for the building are calculated automatically so that all the buildings are separated. Next step the filters were applied for removing the noise that are the unwanted small objects such as cars and other than building objects.

Building count calculated by count each building box and then the area is calculated for each building concerning the Google earth Pro image resolution for available images.

Table 1. Result analysis

Area	TP	FP	FN	Total buildings	Miss factor %	Branching factor %	Building detection %
Samarth Nager	275	146	25	429	0.091	0.531	91.67
Sangita colony	400	129	50	570	0.125	0.323	88.89
Average					0.21	0.17	82.75

9   Conclusion

The result gives a satisfactory result for building detection. The morphological operation performs identifiable with median filter after the OTSU'S Method. This count of the building are will be verified by the technical department of government for accuracy. This also has lacunas as some buildings do still not identify the further research will be carried out for the same.

References

1. Lin, C., Nevatia, R.: Building detection and description from a single intensity image. Comput. Vis. Image Underst. **72**(2), 101–121 (1998). https://doi.org/10.1006/cviu.1998.0724

2. Mayer, H.: Automatic object extraction from aerial imagery—a survey focusing on buildings. Comput. Vis. Image Underst. **74**(2), 138–149 (1999). https://doi.org/10.1006/cviu.1999.0750

3. Wei, Y., Zhao, Z., Song, J.: Urban building extraction from high-resolution satellite panchromatic image using clustering and edge detection. In: Geoscience and Remote Sensing Symposium 2004, IGARSS 2004, Proceedings 2004 IEEE International, vol. 3, pp. 2008–2010. IEEE (2004)

4. Jin, X., Davis, C.H.: Automated building extraction from high-resolution satellite imagery in urban areas using structural, contextual, and spectral information. EURASIP J. Adv. Sig. Process. **2005**(14), 1–11 (2005). https://doi.org/10.1155/ASP.2005.2196

5. Lefèvre, S., Weber, J., Sheeren, D.: Automatic building extraction in VHR images using advanced morphological operators. In: Urban remote sensing joint event 2007, pp. 1–5. IEEE (2007)

6. Huang, X., Zhang, L.: Morphological building/shadow index for building extraction from high-resolution imagery over urban areas. IEEE J. Sel. Top. Appl. Earth Observations Remote Sens. **5**(1), 161–172 (2012). https://doi.org/10.1109/JSTARS.2011.2168195

7. Huang, X., Zhang, L.: A multidirectional and multiscale morphological index for automatic building extraction from multispectral GeoEye-1 imagery. Photogramm. Eng. Remote. Sens. **77**(7), 721–732 (2011). https://doi.org/10.14358/PERS.77.7.721

8. Singh, K.K., Mehrotra, A.: Building extraction from VHR imagery using morphological shadow and building operator. In: Computing for Sustainable Global Development (INDIACom) 2014 International Conference, pp. 463–467. IEEE (2014)

9. Huang, X., Lu, Q., Zhang, L.: A multi-index learning approach for classification of high-resolution remotely sensed images over urban areas. ISPRS J. Photogramm. Remote. Sens. **90**, 36–48 (2014). https://doi.org/10.1016/j.isprsjprs.2014.01.008

10. Huang, X., Han, X., Zhang, L., Gong, J., Liao, W., Benediktsson, J.A.: Generalized differential morphological profiles for remote sensing image classification. IEEE J. Sel. Top. Appl. Earth Observations Remote Sens. **9**(4), 1736–1751 (2016). https://doi.org/10.1109/JSTARS.2016.2524586

11. Pesaresi, M., Benediktsson, J.A.: A new approach for the morphological segmentation of high-resolution satellite imagery. IEEE Trans. Geosci. Remote Sens. **39**(2), 309–320 (2001). https://doi.org/10.1109/36.905239

12. Benediktsson, J.A., Pesaresi, M., Amason, K.: Classification and feature extraction for remote sensing images from urban areas based on morphological transformations. IEEE Trans. Geosci. Remote Sens. **41**(9), 1940–1949 (2003). https://doi.org/10.1109/TGRS.2003.814625

13. Singh, D., Maurya, R., Shukla, A.S., Sharma, M.K., Gupta, P.R.: Building extraction from very high resolution multispectral images using NDVI based segmentation and morphological operators. In: Engineering and Systems (SCES), 2012 Students Conference, pp. 1–5. IEEE (2012)

14. San, D.K., Turker, M.: Building extraction from high resolution satellite images using Hough transform. Remote Sensing and Spatial Information Science, vol. XXXVIII, Part 8, Kyoto Japan (2010)

15. Huang, X., Zhang, L.: An SVM ensemble approach combining spectral, structural, and semantic features for the classification of high-resolution remotely sensed imagery. IEEE Trans. Geosci. Remote Sens. **51**(1), 257–272 (2013). https://doi.org/10.1109/TGRS.2012.2202912

16. Dey, V., Zhang, Y., Zhong, M.: Building detection from pan-sharpened GeoEye-1 satellite imagery using context-based multi-level image segmentation. In: Image and Data Fusion (ISIDF), 2011 International Symposium, pp. 1–4. IEEE (2011)

17. Zhan, Q., Molenaar, M., Tempfli, K., Shi, W.: Quality assessment for geospatial objects derived from remotely sensed data. Int. J. Remote Sens. **26**(14), 2953–2974 (2005). https://doi.org/10.1080/01431160500057764

18. Shan, J., Lee, S.D.: Quality of building extraction from IKONOS imagery. J. Surv. Eng. **131**(1), 27–32 (2005). https://doi.org/10.1061/(ASCE)0733-9453(2005)131:1(27)
19. Sreedhar, K., Panlal, B.: Enhancement Of images using morphological transformations. Int. J. Comput. Sci. Inf. Technol. (IJCSIT) **4**(1), 33–50 (2012)
20. Go"ksel, C., Kaya, S., Musaog"lu, N.: Satellite data use for change information: A case study for Terkos water basin ˙Istanbul, 21. In: EARSeL Symposium, proceedings Paris, France, pp. 299–302 (2001)
21. Jensen, J.R.: Introductory Digital Image Processing a Remote Sensing Perspective, p. 318. Prentice-Hall, Upper Saddle River (1996)
22. Jha, C.S., Unni, N.V.M.: Digital change detection of forest conversion of a dry tropical Indian forest region. Int. J. Remote Sens. **15**(13), 2543–2552 (1994)
23. Rafael, C., Gonzalez, R.E., Woods, S.L.: Eddins. Digital Image Processing Using MATLAB, Pearson Prentice Hall, New Jersey (2004)
24. Lau, S.: Global image enhancement using local information. Electron. Lett. **30**, 122–123 (1994)
25. Wenqing, L.J.L., Jianzhuang, L.: The automatic threshold of gray-level pictures via two-dimensional Otsu method. Acta Autom. Sin. **1**, 15 (1993)
26. Gong, J., Li, L., Chen, W.: Fast recursive algorithms for two-dimensional thresholding. Pattern Recogn. **31**(3), 295–300 (1998)
27. Sahoo, P.K., Soltani, S., Wong, A.K.C., Chen, Y.: A survey of thresholding techniques. Comput. Vis. Graph. Image Process. **41**, 233–260 (1988)
28. Kirwale, K.S., Kawathekar, S.S., Deshmukh, R.R.: Building detection from High resolution images using morphological operation. IOSR J. Comput. Eng. **19**(6), 37–41 (2017)

A Liver Segmentation Algorithm
with Interactive Error Correction
for Abdominal CT Images: A Preliminary Study

P. Vaidehi Nayantara[1], Surekha Kamath[1(✉)], K. N. Manjunath[2], and K. V. Rajagopal[3]

[1] Department of Instrumentation and Control Engineering, Manipal Institute of Technology, Manipal Academy of Higher Education, Manipal, Karnataka 576104, India
surekha.kamath@manipal.edu

[2] Department of Computer Science and Engineering, Manipal Institute of Technology, Manipal Academy of Higher Education, Manipal, Karnataka 576104, India

[3] Department of Radiodiagnosis and Imaging, Kasturba Medical College, Manipal Academy of Higher Education, Manipal, Karnataka 576104, India

Abstract. An automatic method for segmenting the liver from the portal venous phase of abdominal CT images using the K-Means clustering method is described in this paper. We have incorporated an interactive technique for correcting the errors in the liver segmentation results using power law transformation. The proposed method was validated on abdominal CT volumes of fifteen patients obtained from Kasturba Medical College, Manipal. The average values of the various standard evaluation metrics obtained are as follows: Dice coefficient = 0.9361, Jaccard index = 0.8805, volumetric overlap error = 0.1195, absolute volume difference = 4.048%, average symmetric surface distance = 1.7282 mm and maximum symmetric surface distance = 38.039 mm. The quantitative and qualitative results obtained in our preliminary work show that the K-Means clustering technique along with power law transformation is effective in producing good liver segmentation outputs. As future work, we will attempt to automate the power law transformation technique.

Keywords: Preprocessing · Liver segmentation · Power law transformation · K-means clustering

1 Introduction

Segmentation of the liver from the abdominal Computed Tomography (CT) image is essential for the computer based diagnosis of hepatic diseases and liver surgery planning [1, 2]. But the pixel intensity of liver parenchyma is very similar to its adjacent organs like stomach, heart, kidney, etc., making it difficult to extract only the liver region from the abdominal image [3, 4]. Hence portions of these adjacent organs also get segmented along with the liver. Another factor that complicates the segmentation process is the presence of large peripheral or border tumors in the liver. These tumors often get excluded from the segmented liver region yielding impractical segmentation results.

© IFIP International Federation for Information Processing 2021
Published by Springer Nature Switzerland AG 2021
V. Krishnamurthy et al. (Eds.): ICCIDS 2021, IFIP AICT 611, pp. 132–140, 2021.
https://doi.org/10.1007/978-3-030-92600-7_13

Several papers have been published on liver segmentation [5–12]. Most of the research focused on automating the segmentation process with no provision for manual intervention to identify and correct the erroneously segmented slices in the CT dataset of the patient. Automation is essential but may not always be practical as there may be errors in the segmented output. These errors must be corrected to get accurate outcomes in computer-assisted systems used for hepatic disease diagnosis and surgery. Accuracy is foremost in medical applications and no algorithm can guarantee to always produce accurate results. Hence it is essential to incorporate some method for interactive correction of the erroneously segmented results.

We have attempted to segment the liver from the images obtained in the portal venous CT phase using the K-Means clustering algorithm. Power law transformation interactively corrected the erroneously segmented images. Our preliminary work has shown promising results both qualitatively and quantitatively. There is no work published on liver segmentation that uses power law transformation for error correction to the best of our knowledge. The paper is structured as follows: Sect. 2 describes the methods adopted in the proposed work, Sect. 3 presents the results and discussion; conclusion and insights into the future work are provided in Sect. 4.

2 Methods

The different steps in the proposed liver segmentation method are depicted in the flow diagram shown in Fig. 1. It mainly consists of four stages: preprocessing, K-Means clustering based liver segmentation, postprocessing and an error correction block based on power law transformation. Each of these stages are detailed in the following subsections. The results obtained at the different stages of the algorithm for a particular input image are illustrated in Fig. 2.

2.1 Preprocessing

The images considered in this work are in Digital Imaging and Communications in Medicine (DICOM) format. The images are first converted from CT numbers to Hounsfield Units (HU) using Eq. 1 [13],

$$HU = CTnumber \times Rescale\_Slope + Rescale\_Intercept \qquad (1)$$

where *CTnumber* is the stored value at each image point and the values of *Rescale_Slope* and *Rescale_Intercept* are obtained from the DICOM file itself using the respective tags.

The pixel values obtained after the above transformation are windowed to 0–255 range according to the pseudocode given below for two reasons: (i) to make them suitable for display, (ii) since images in HUs did not give satisfactory segmentation results. The pseudocode for the same adapted from [14] is shown below.

n = number of output gray levels (i.e., 256)
w = Window width – 1.0
c = Window center – 0.5
if (HU <= c−0.5 * w), then Windowed value = 0

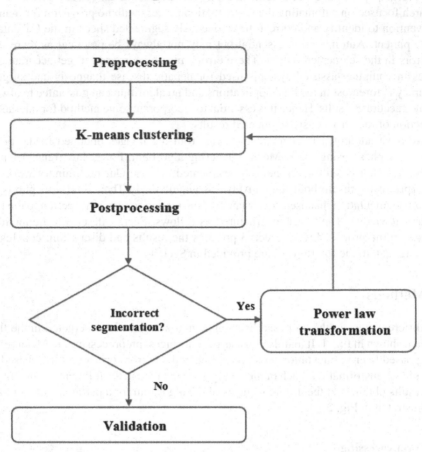

Fig. 1. Flow diagram of the proposed liver segmentation algorithm.

else if (HU > c + 0.5 * w), then Windowed value = n−1
else Windowed value = ((HU-c)/w + 0.5) * (n−1)

Here the values of Window width and Window center are obtained from the DICOM file. The image obtained after these operations is shown in Fig. 2(a).

To remove the ribs, spine, etc. (referred to as white region in this paper) from the windowed CT image, the pixels with intensity above 250 are mapped to 0. Then unsharp masking is performed to sharpen the image. Unsharp masking is an image enhancement technique in which sharp details like edges are accentuated in two steps. First, the blurred version of the image (low pass filtered) is subtracted from the original image to obtain the sharp details. Then these details are scaled and added back to the original image to get the sharpened image [15]. Mathematically, it is defined as,

$$f_{sh}(x, y) = f(x, y) + c * (f(x, y) - f_{bl}(x, y)) \tag{2}$$

where $f_{sh}(x, y)$ is the sharpened image, $f(x, y)$ is the original image, $f_{bl}(x, y)$ is the blurred version of the original image and c is the scaling factor. For blurring the image, the Gaussian filter with standard deviation, $\sigma = 1$ was used, and the scaling factor, c chosen, was 0.8. The image thus obtained is shown in Fig. 2(b).

2.2   K-Means Clustering and Postprocessing

The liver segmentation is done by the K-Means clustering algorithm to segment the CT volume into three clusters. The three clusters correspond to: (i) the background region, (ii) the liver and other structures with similar gray level intensity and (iii) the non-hepatic areas. The initial centroids were selected using the k-means++ algorithm. The algorithm consists of two stages. The first stage calculates the K centroids and the second stage assigns each point to the cluster with the nearest centroid. Once the grouping is done, it recalculates the new centroid for each cluster by averaging the data points. This process is repeated till there is no change in the centroids (less than 0.0001). The image corresponding to the liver cluster is shown in Fig. 2(c).

Several morphological operations like opening, closing, hole filling and largest connected component extraction are performed in the postprocessing stage to remove imperfections like holes, bridges, etc. and retain only the liver. After performing these operations, we get the binary mask of liver shown in Fig. 2(d). The liver image shown in Fig. 2(e) is obtained by multiplying the binary mask of liver with the windowed image of Fig. 2(a).

2.3   Interactive Error Correction

Power law transformation was used for correcting the segmentation errors that resulted after segmenting some of the complex datasets. Power law transformation is a basic gray-level transformation function that is used for performing image enhancement. It is mainly used for contrast manipulation and gamma correction [16].

Mathematically, it is defined as,

$$s = cr^\gamma \tag{3}$$

where c and γ are positive constants; r and s are values of pixels before and after enhancement.

The incorrectly segmented images were corrected interactively by varying the value of γ in Eq. 3 to adjust the contrast between the liver and its adjacent organs or between the liver and its peripheral tumors. The power law transformed images were again subjected to K-Means clustering and postprocessing in order to get the correctly segmented output. Figure 3 illustrates how the power law transformation method helps in correcting the incorrectly segmented results.

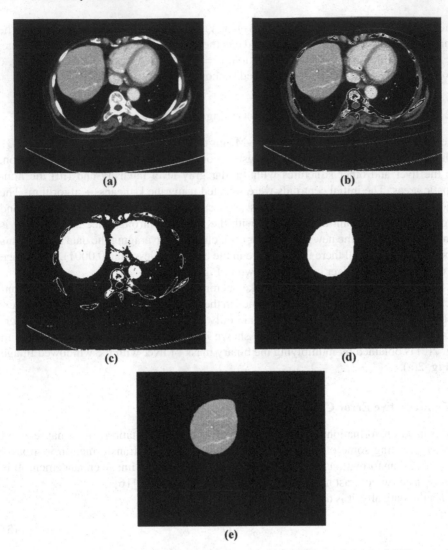

Fig. 2. Results of the proposed liver segmentation algorithm. (a) Input abdominal CT image after windowing (b) Output after white region removal and unsharp masking (c) Output of K-Means clustering based segmentation (d) Binary mask of the liver after postprocessing (e) Segmented liver.

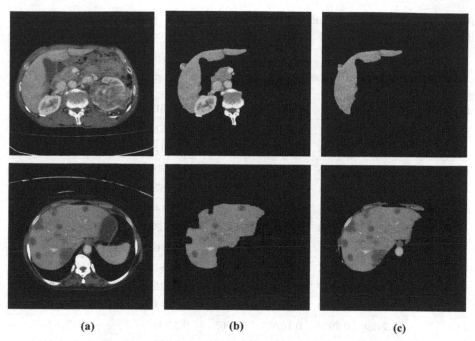

(a)   **(b)**   **(c)**

Fig. 3. Correction of incorrectly segmented outputs using power law transformation. (a) Input abdominal CT image (b) Incorrectly segmented output (c) Segmented output after application of power law transformation.

3 Results and Discussion

We implemented the proposed liver segmentation algorithm in MATLAB R2020b. The evaluation of the algorithm was done both qualitatively and quantitatively. The qualitative analysis and the generation of the ground truth mask were done under the supervision of a senior radiologist with more than twenty years of experience. The ground truth was generated using the ITK-SNAP tool [17].

Six metrics, namely Dice Coefficient (DC), Jaccard Index (JI), Volumetric Overlap Error (VOE), Absolute Volume Difference (AVD), Average Symmetric Surface Distance (ASSD) and Maximum Symmetric Surface Distance (MSSD) were used for evaluating the quality of segmentation. The different metrics compute the segmentation accuracy from different perspectives viz. volume overlap and surface distance perspectives. DC and JI indicate the amount of overlap between the segmented and ground truth volumes. AVD indicates the difference in the size of the two volumes. The ASSD and MSSD indicate the closeness of the surface voxels of the two volumes. For DC and JI, values closer to 1 indicate better segmentation, whereas, for the rest, values nearer to 0 indicate better results [18, 19].

The quantitative results in terms of the above mentioned six metrics are given in Table 1. The CT volumes (portal venous phase) of fifteen different patients obtained from Kasturba Medical College (KMC), Manipal, were used for segmentation evaluation. Twenty consecutive liver slices were selected from each of the fifteen patient CT volumes for evaluating the proposed algorithm. Hence, in Table 1, each dataset corresponds to the abdominal CT volume of a patient comprising twenty consecutive liver slices.

Table 1. Quantitative evaluation results of the proposed algorithm on 15 patient datasets.

Dataset no.	DC	JI	VOE	AVD (in %)	ASSD (in mm)	MSSD (in mm)
1	0.9253	0.8611	0.1389	4.2690	1.1670	132.4307
2	0.8989	0.8164	0.1836	12.6408	3.8375	28.4008
3	0.9336	0.8755	0.1245	5.5231	0.7301	18.5428
4	0.9459	0.8974	0.1026	2.3014	2.6676	42.2225
5	0.9519	0.9083	0.0917	4.2044	2.4149	54.7608
6	0.9392	0.8854	0.1146	5.8780	1.7038	16.5062
7	0.9432	0.8926	0.1074	5.3560	1.7164	18.8641
8	0.9750	0.9512	0.0488	1.7162	1.3819	11.7804
9	0.9222	0.8557	0.1443	2.3571	1.0467	36.2227
10	0.9347	0.8774	0.1226	0.5670	0.6148	7.5158
11	0.9429	0.8920	0.1080	3.6765	2.7004	40.7182
12	0.9290	0.8673	0.1327	4.6857	0.8551	44.4336
13	0.9482	0.9016	0.0984	3.7210	1.7413	40.2279
14	0.9436	0.8933	0.1067	1.6448	1.6548	29.1049
15	0.9081	0.8316	0.1684	2.1789	1.6906	48.8543
Average	**0.9361**	**0.8805**	**0.1195**	**4.0479**	**1.7282**	**38.0390**

The average values of the standard metrics achieved by our algorithm (DC = 0.9361, JI = 0.8805, VOE = 0.1195, AVD = 4.0479%, ASSD = 1.7282 mm and MSSD = 38.039 mm) were promising. The best values for DC, JI and VOE were 0.9750, 0.9512 and 0.0488, respectively, corresponding to Dataset 8 and the best values for AVD, ASSD and MSSD were 0.5670%, 0.6148 mm and 7.5158 mm, respectively corresponding to Dataset 10. These results show that the algorithm is effective in segmenting the liver from the abdominal CT volume. Specifically, it has shown the effectiveness of the power law transformation technique in correcting the segmentation errors. The main pitfall of the method is the need to input the gamma value manually. In the future, we intend to formulate a method for automatically computing this value based on the attributes of the input image.

4   Conclusion and Future Work

We have performed a preliminary study to segment the liver from the abdominal CT volume using the K-Means clustering technique. An interactive error correction with power law transformation method was also incorporated in our algorithm. We have achieved good qualitative and quantitative results with some of the challenging datasets. One drawback of the algorithm is that the error correction technique is semiautomatic. As future work, we will formulate a method for automatically computing the gamma value for the power law transformation method based on the contrast, intensity values, histogram and other factors associated with the input image.

Acknowledgments. The work is supported by KStePS, DST, Government of Karnataka, India. The authors are grateful to Manipal Institute of Technology, MAHE, Manipal for providing the facilities to carry out the research and Kasturba Medical College, Manipal, for providing the patient data.

References

1. Campadelli, P., Casiraghi, E., Esposito, A.: Liver segmentation from computed tomography scans: a survey and a new algorithm. Artif. Intell. Med. **45**(2–3), 185–196 (2009). https://doi.org/10.1016/j.artmed.2008.07.020
2. Lim, S.-J., Jeong, Y.-Y., Ho, Y.-S.: Automatic liver segmentation for volume meas-urement in CT Images. J. Vis. Commun. Image Represent. **17**(4), 860–875 (2006). https://doi.org/10.1016/j.jvcir.2005.07.001
3. Moghbel, M., Mashohor, S., Mahmud, R., Saripan, M.I.B.: Review of liver segmentation and computer assisted detection/diagnosis methods in computed tomography. Artif. Intell. Rev. **50**(4), 497–537 (2017). https://doi.org/10.1007/s10462-017-9550-x
4. Gotra, A., et al.: Liver segmentation: indications, techniques and future directions. Insights Imaging **8**(4), 377–392 (2017). https://doi.org/10.1007/s13244-017-0558-1
5. Siri, S.K., Latte, M.V.: Universal liver extraction algorithm: an improved Chan–vese model. J. Intell. Syst. **29**(1), 237–250 (2020)
6. Xu, L., Zhu, Y., Zhang, Y., Yang, H.: Liver segmentation based on region growing and level set active contour model with new signed pressure force function. Optik (Stuttg.) **202**(July), 2019 (2020). https://doi.org/10.1016/j.ijleo.2019.163705
7. Satpute, N., Gómez-Luna, J., Olivares, J.: Accelerating Chan-Vese model with cross-modality guided contrast enhancement for liver segmentation. Comput. Biol. Med. **124**, 103930 (2020). https://doi.org/10.1016/j.compbiomed.2020.103930
8. Li, Y., et al.: Liver segmentation from abdominal CT volumes based on level set and sparse shape composition. Comput. Methods Programs Biomed. **195**, 105533 (2020). https://doi.org/10.1016/j.cmpb.2020.105533
9. Danilov, A., Yurova, A.: Automated segmentation of abdominal organs from contrast-enhanced computed tomography using analysis of texture features. Int. J. Numer. Method. Bbiomed. Eng. **36**(4), 1–14 (2020). https://doi.org/10.1002/cnm.3309
10. Muthuswamy, J.: Extraction and classification of liver abnormality based on neutrosophic and SVM classifier. In: Pati, B., Panigrahi, C.R., Misra, S., Pujari, A.K., Bakshi, S. (eds.) Progress in Advanced Computing and Intelligent Engineering. AISC, vol. 713, pp. 269–279. Springer, Singapore (2019). https://doi.org/10.1007/978-981-13-1708-8_25
11. Lu, X., Xie, Q., Zha, Y., Wang, D.: Fully automatic liver segmentation combining multi-dimensional graph cut with shape information in 3D CT images. Sci. Rep. **8**(1), 10700 (2018). https://doi.org/10.1038/s41598-018-28787-y
12. Kumar, S.S., Moni, R.S., Rajeesh, J.: Automatic liver and lesion segmentation: a primary step in diagnosis of liver diseases. Signal, Image Video Process. **7**(1), 163–172 (2013). https://doi.org/10.1007/s11760-011-0223-y

13. "DICOM Documentation- Modality Specific Modules." http://dicom.nema.org/medical/dicom/current/output/chtml/part03/sect_C.8.15.3.10.html. Accessed 20 Jan 2021
14. "DICOM Documentation – Look Up Tables and Presentation States." http://dicom.nema.org/medical/dicom/current/output/chtml/part03/sect_C.11.2.html#sect_C.11.2.1.2.1. Accessed 20 Jan 2021
15. Jain, A.K.: Fundamentals of Digital Image Processing, Prentice Hall, Englewood. Cliffs (1989)
16. Gonzalez, R., Woods, R.: Digital Image Processing, 3rd edn. Prentice-Hall, Inc., Englewood. Cliffs (2006)
17. Yushkevich, P.A., Gao, Y., Gerig, G.: ITK-SNAP: an interactive tool for semi-automatic segmentation of multi-modality biomedical images. In: 2016 38th Annual International Conference of the IEEE Engineering in Medicine and Biology Society (EMBC), pp. 3342–3345 (2016)
18. Taha, A.A., Hanbury, A.: Metrics for evaluating 3D medical image segmentation: analysis, selection, and tool. BMC Med. Imaging **15**, 29 (2015). https://doi.org/10.1186/s12880-015-0068-x
19. Yeghiazaryan, V., Voiculescu, I.: Family of boundary overlap metrics for the evaluation of medical image segmentation. J. Med. Imaging (Bellingham, Wash.), **5**(1), 15006 (2018). https://doi.org/10.1117/1.JMI.5.1.015006

Pixel Based Adversarial Attacks
on Convolutional Neural Network Models

Kavitha Srinivasan[(✉)] [iD], Priyadarshini Jello Raveendran, Varun Suresh,
and Nithya Rathna Anna Sundaram

Department of Computer Science and Engineering, Sri Sivasubramaniya Nadar College
of Engineering, Kalavakkam 603 110, India
kavithas@ssn.edu.in, {priyadarshini15074,varun15122,
nithyarathna15067}@cse.ssn.edu.in

Abstract. Deep Neural Networks (DNN) has found their applications in the real
time, for example, facial recognition for security in ATMs and self-driving cars.
A major security threat to DNN is through adversarial attacks. An adversarial
sample is an image that has been changed in such a way that it is imperceptible
to human eye but causes the image to be misclassified by a Convolutional Neural
Networks (CNN). The objective of this research work is to devise pixel based
algorithms for adversarial attacks on images. For validating the algorithms, untargeted attack is performed on MNIST and CIFAR-10 dataset using techniques such
as edge detection, Gradient weighted Class Activation Mapping (GRAD-CAM)
and noise addition whereas targeted attack is performed on MNIST dataset using
Saliency maps. These adversarial images thus generated are then passed to a CNN
model and the misclassification results are analyzed. From the analysis, it has
been inferred that it is easier to fool CNNs using untargeted attacks than the targeted attacks. Also, grayscale images (MNIST) are preferred to generate robust
adversarial examples compared to colored images (CIFAR-10).

Keywords: Deep Neural Networks · Adversarial attacks · Convolutional Neural
Network Models · Gradient weighted Class Activation Mapping · Edge detection ·
Noise addition · Saliency maps

1 Introduction

Deep Learning is finding its use in many applications nowadays. Unsolved problems
of the Machine Learning and Artificial Intelligence are being solved by Deep Learning
techniques. As a result, it is currently being used in a variety of real-world application
such as human face recognition, image analysis and self-driving cars. These applications
lead to large volume of information and computations. Face recognition tool or software,
efficient enough to be used in ATMs and for unlocking phones, have been developed.
Some of the medical systems developed for analysis and disease detection performs
better than human experts of that specific field. Self driving cars are a distinguishable
application where human drivers are no longer required.

© IFIP International Federation for Information Processing 2021
Published by Springer Nature Switzerland AG 2021
V. Krishnamurthy et al. (Eds.): ICCIDS 2021, IFIP AICT 611, pp. 141–155, 2021.
https://doi.org/10.1007/978-3-030-92600-7_14

The latest research by Google Brain has shown that any machine learning classifier could be modified to result a wrong outcome. The attackers can showcase their skills by changing the result as per their requirement. This scenario affected the real time systems in banks, ATMs, facial recognition on laptops and self-driving cars, which are developed by artificial intelligence and many of them are decisive for a secured life.

For example, if an attacker aims to create adversarial attack on direction signs, then the self-driving car may interpret in a wrong way and can take unwanted actions. This might result in a major road accident causing severe damage. An adversarial attack involves slightly changing an image in such a way that the modifications are indistinguishable to the human eyes. The changed image is called as adversarial image, which results a misclassification when submitted to the classifier. The two types of adversarial attacks are targeted attack and untargeted attack. In a targeted attack, the attacker dissimulates to get the image classified as a specific target class, which is different from the class of the original image. The objective of untargeted attacks is to make the model to return a wrong prediction as outcome using adversarial image.

This research work is proposed to create untargeted attacks using techniques such as edge detection, GRAD-CAM and noise application and targeted attacks using Saliency maps. Edge detection involves detecting the edge pixels using Canny library and altering their intensity values. Grad-CAM takes the gradients from the final convolutional layer to generate a map that helps determine the essential regions in an image. Saliency maps are generated by computing the gradient of output image with respect to input image. These algorithms are an evidence to understand how the output value changes with respect to a small modification in input image pixels. These adversarial images thus generated are then passed to a CNN model and the results are compared and analyzed.

2 Related Work

Adversarial samples are inputs to machine learning models that an attacker has deliberately designed a slight perturbation to cause the model to misclassify an image. The fascinating properties of neural networks perturbations are applied to an image that caused the network to misclassify it (Szegedy et al. 2014). These perturbations were found by maximizing the network's prediction error. Attacks and defenses for deep learning model are defined as adversarial examples to be invisible to human eyes that could easily fool deep neural networks is elaborated in (Yuan et al. 2019). The paper summarized recent discoveries on adversarial examples, methods for generating adversarial examples and also proposed categorization of those methods.

One Pixel attack (Su et al. 2019) is carried out by changing only one pixel with differential evolution in a scenario where the only information available is the probability labels. The attack was tested on pre-trained models on CIFAR-10 and Alexnet mode trained on Imagenet. Zeroth Order Optimization (Chen et al. 2017) based attacks directly estimate the gradients of the targeted DNN for generating adversarial examples.

The one-step gradient-based approach namely FGSM, finds an adversarial example by maximizing the loss function (Goodfellow et al. 2015). Iterative methods run FGSM multiple times with a small step size α and named as I-FGSM (Kurakin et al. 2015). MI-FGSM is a method for speeding up gradient descent algorithms by accumulating a velocity vector in the direction of loss function across iterations (Dong et al. 2018). These iterative based attacks were trained on Imagenet and tested on Inception network.

Carlini and Wagner attacks are adapted to three distance metrics L_0, L_2 and L_∞ (Carlini and Wagner 2017). The target is to find δ that minimizes $D(x, x + \delta)$, where x is the given image. That is, to find some small change δ that can be made to an image x that will change its classification, however the result would still be a valid image. In Table 1, the summary of existing approaches is discussed.

In Adversarial Attacks and Defenses: A Survey by (Chakraborty et al. 2018), explained about different threat models, existing attack algorithms and its countermeasures. The paper compares the efficiency and limitations of the various attacks and countermeasures that were selected for examination.

In Practical Adversarial Attack against Object Detector by (Zhao et al. 2018), discussed two different attack algorithms against object detectors in realistic situations such as autonomous driving cars are provided. YOLO V3 was the object detector upon which the attacks were tested on by changing various factors such as distance, illumination, angles etc.

The research papers mentioned have created an algorithm that makes subtle changes in the images causing neural networks to misclassify the adversarial images. This establishes the fact that neural networks are vulnerable to adversarial changes. The goal of this project is to create similar yet simpler attack algorithms using pixel modification techniques. For instance, by causing changes to the background pixels without affecting the pixels that affect the shape of the number, misclassifications were caused. This shows that CNNs do not view images as humans do and are vulnerable to adversarial attacks. Attack algorithms were hence created, applied on images from MNIST and CIFAR-10 datasets.

Table 1. Summary of existing attack methods

Method	Type	Dataset	Model used
One pixel attack	Untargeted	CIFAR-10 ImageNet	VGG AlexNet
Zeroth order optimisation	Targeted Untargeted	MNIST CIFAR-10 ImageNet	C&W Framework C&W Framework Inception v3
Fast gradient sign method (FGSM)	Targeted	MNIST ImageNet	Softmax Classifier Inception v3
Iterative-FGSM	Targeted	ImageNet	Inception v3
Momentum iterative – FGSM	Targeted	ImageNet	Inception v3
C&W attack	Targeted	MNIST CIFAR-10 ImageNet	C&W Framework C&W Framework AlexNet

3 Proposed System

The proposed system aims to create untargeted and targeted attacks using techniques such as edge detection, GRAD-CAM, noise addition and saliency maps. The proposed algorithms (adversarial image creation) for two datasets are given in Fig. 1.

Input: 28x28 black and white MNIST
              32x32 colored CIFAR-10 image
Output: Attacked image
1. *Untargeted Attacks on MNIST*
   a. Modification of background pixels o
   b. Modification of edge pixels
   c. Modification of edge and background pixels
   d. Gaussian noise addition and modification of edge pixels
2. *Untargeted Attacks on CIFAR-10*
   a. Modification of edge pixels
   b. Heatmap generation and modification of edge pixels
3. *Targeted Attacks on MNIST*
   a. Altering effective pixels with fixed values
   b. Altering effective pixels using AES
   c. Altering maximum occurring pixels

Fig. 1. Workflow of proposed system

3.1 Untargeted Attacks

An untargeted attack aims to affect the image such that the model misclassifies the image. The untargeted attack is performed on MNIST and CIFAR-10 dataset using various techniques as mentioned in Fig. 1.

The methods used to perform untargeted attacks are explained below:

Canny Edge Detection Algorithm: The canny edge detector is a multistage edge detection algorithm. It has 4 stages such as preprocessing, calculating gradients, non-maximum suppression and thresholding with hysteresis. The two key parameters of the algorithm are - an upper threshold and a lower threshold. The upper threshold is used to mark edges that are definitely edges. The lower threshold is to find faint pixels that are actually a part of an edge.

Heat Map: A heat map is a visual representation of data where the individual values contained in a matrix are represented as colors. It helps in decision making by highlighting areas of greater attention. Areas of high activity are represented using bright colors while the areas of low activity are represented using darker colors.

AES Algorithm: The Advanced Encryption Standard (AES) is a symmetric block cipher. It is the most popular and widely used encryption algorithm because it is six times faster and has a smaller key size than Data Encryption Standard (DES). It contains four phases such as add round key, substitute bytes, shift rows and mix columns to perform encryption.

3.1.1 MNIST

In the MNIST dataset, there are two sets of pixels. One set contributes to the background while the other contributes to the shape of the number. The simplest change that could be caused to one such image is altering the intensity of the background pixels.

After the attack caused misclassification, the next step was to verify there would be a similar effect upon altering the pixels that affect the shape of the number. Thus, the pixels contributing the edge of the numbers (along with their surrounding pixels) were detected and their intensity was altered using AES algorithm. As the misclassification did not match up to the background change, the two methods were combined.

Modification of Background Pixels: The background color of the image is modified by adding a constant value.

Algorithm 1: Attack algorithm for modifying background pixels

Input: 2D black and white image of handwritten digits from 0 to 9 of dimensions 28x28 (MNIST)

Output: Attacked image

    1:   **function** BACKGROUND PIXEL MODIFICATION(*Image*)

    2:        Identify pixels with intensity zero

    3:        **for all** pixels i in identified pixels **do**

    4:            Add a constant value to the intensity

    5:        **end for**

    6:        **return** *Image*

    7: **end function**

Edge Detection: Edge pixels are identified using Canny library and the intensity of these pixels are modified using AES algorithm.

Algorithm 2: Attack algorithm for modifying edge pixels

Input: 2D black and white image of handwritten digits from 0 to 9 of dimensions 28x28 (MNIST)

Output: Attacked image

  1:   **function** EDGE PIXEL MODIFICATION(*Image*)

  2:           Identify the edge pixels using Canny library

  3:           Add the pixels to the left, right, top and bottom of identified pixel to the list

  4:           **for all** pixels i in identified pixels do

  5:                 Apply AES algorithm on the intensity value

  6:                 Perform mod 256 on the obtained value

  7:                 Set this value as the new intensity of the pixel i

  8:           **end for**

  9:           **return** *Image*

10: **end function**

Combination of Edge Detection and Background Modification: Intensity of edge pixels and background pixels are modified using AES algorithm in which two different nonces are set for edge pixels and background pixels.

Algorithm 3: Attack algorithm for modifying edge and background pixels

Input: 2D black and white image of handwritten digits from 0 to 9 of dimensions 28x28 (MNIST)

Output: Attacked image

  1:   **function** EDGE & BACKGROUND MODIFICATION(*Image*)

  2:           Identify pixels in Image with intensity zero

  3:           Identify edge pixels in Image using Canny Algorithm

  4:           **for all** pixels i in identified pixels do

  5:                 If pixel i is an edge pixel set nonce as '!#!9'

  6:                 If pixel i is a background pixel set nonce as '!#!#'

  7:                 Apply AES algorithm on the intensity value

  8:                 Perform mod 256

  9:                 Set this value as the new intensity of the pixel i

10:         **end for**

11:         **return** *Image*

12: **end function**

Addition of Noise and Edge Detection: Gaussian noise is applied to the image and then AES algorithm is applied to modify the intensity of those edge pixels.

Algorithm 4: Attack algorithm for addition of gaussian noise and modifying edge pixels

Input: 2D black and white image of handwritten digits from 0 to 9 of dimensions 28x28 (MNIST)
Output: Attacked image

  1:   **function** NOISE ADDITION(*Image*)
  2:        Add Gaussian noise to the image using random noise function
  3:        Determine the edge pixels using Canny library
  4:        Add the pixels to the bottom-left, top-left, bottom-right and top-right of identified pixel to the   list
  5:        **for all** pixels i in identified pixels **do**
  6:            Apply AES algorithm on the intensity value with nonce '!#!%'
  7:            Perform mod 256 on the obtained value
  8:            Set this value as the new intensity of the pixel i
  9:        **end for**
10:        **return** *Image*
11: **end function**

3.1.2 CIFAR-10

The images from CIFAR-10 have three channels, red-blue-green. As colored images do not have differentiated set of pixels, it did not make sense to apply background modification on this dataset. Similarly noise application could cause random spots of color and hence was not tested on CIFAR-10.

While edge pixels contribute to the shape of the number in MNIST, they contribute to the object or some object in the background in CIFAR-10. For instance, an image of a horse in front of a mountain causes the edge of the mountain to be detected as well. Thus, the effect of the attack is less prominent in CIFAR-10. The pixel intensities have to be altered for each channel. Finally, we determined the important regions in an image by using Gradient-weighted Class Activation Mapping (Grad-CAM). Based on the heatmap generated from the image, the pixels contributing to the important regions were modified.

Edge Detection: Edge pixels are identified using Canny library and the intensity of these pixels are modified using AES algorithm.

Algorithm 5: Attack algorithm for modifying edge pixels

Input: 2D color image of dimensions 32x32 (CIFAR-10)

Output: Attacked image

1: **function** EDGE PIXEL MODIFICATION(*Image*)

2:          Identify edge pixels in Image using Canny Algorithm

3:          **for all** pixels i in identified pixels do

4:                  Set nonce to 'wxyy' for the blue channel

5:                  Set nonce to 'wxyz' for the green channel

6:                  Set nonce to 'wxyw' for the red channel

7:                  Apply AES algorithm on the channel intensities and perform mod 256

8:                  Set this as the new intensity of pixel i

9:          **end for**

10:          **return** *Image*

11: **end function**

Combination of Heat Map Generation and Edge Detection: Identifying significant pixels using a heat map and modifying them using AES algorithm. If heatmap is not generated, edge pixels are detected and AES is applied to those images.

Algorithm 6: Attack algorithm for modifying significant pixels using a heatmap and modifying edge pixels for others

Input: 2D color image of dimensions 32x32 (CIFAR-10)

Output: Attacked image

1: **function** HEATMAP(Image)

2:          Generate heatmap of the given image

3:          If heatmap is generated, goto step 10

4:          If heatmap is not generated, determine edge pixels using Canny library

5:          **for all** pixels i in identified pixels **do**

6:                  Apply AES algorithm on the intensity value with nonce 'z123'

7:                  Set this value as the new intensity of the pixel i

8:          **end for**

9:          **goto** step 15

10:          Identify the pixels whose channel with maximum intensity is red

11:          **for all** pixels i in identified pixels **do**

12:                  Apply AES algorithm on the intensity value with nonce 'z123'

13:                  Set this value as the new intensity of the pixel i

14:          **end for**

15:          **return** Image

16: **end function**

3.2 Targeted Attacks

In a targeted attack, the attacker aims to get the image classified as a specific target class that is different from the class of the original image. The attack has been applied on a subset of the MNIST dataset. The model is trained on the images of numbers 0, 1, 2 and 7. The attacks applied on the images are Altering effective pixels using fixed values and Altering pixels with maximum occurrence.

The untargeted attacks mentioned in the previous section caused changes to pixel intensities and caused misclassification. This section aims at causing targeted misclassification. Thus, it is important to determine the pixels considered important for a given class. To determine significant pixels for a given class, a saliency map is generated.

The first attack was to reduce the intensity of the pixels that contribute to the current class while increasing the intensity of the pixels contributing to the target class. Though this caused effective misclassification, the changes were visible to a human eye. To prevent this, the pixels were grouped based on their saliency values. The set with the highest occurrence was alone modified. Though the number of misclassification was reduced, the changes caused to the image were less prominent.

3.2.1 Altering Effective Pixels Using Fixed Values
The intensity of effective pixels determined from saliency maps are changed using fixed values, causing the image to be misclassified as belonging to the class chosen by the attacker.

Algorithm 7: Attack algorithm for altering effective pixels with fixed values

Input: 2D black and white image of a handwritten number of 0, 1, 2 or 7 of dimensions 28x28 (MNIST), Target class
Output: Attacked Image

   1:   **function** ALTERING FIXED(Image,Target)
   2:          Change the activation function of the Models last from softmax to linear
   3:          Pick an image T from Target class
   4:          Create a saliency map for both Image and T
   5:          Change the intensity to 0 in Image, for positions in Image with the saliency greater than or equal to 0.1
   6:          Change the intensity to 255 in the Image, for positions in T with the saliency greater than or equal to 0.2
   7:          **return** Image
   8: **end function**

3.2.2 Altering Pixels with Maximum Occurrence
The intensity of maximum occurring pixels determined from saliency maps are changed. This causes the image to be misclassified as belonging to the class chosen by the attacker.

Algorithm 8: Attack algorithm for altering maximum occurring values

Input: 2D black and white image of a handwritten number of 0, 1, 2 or 7 of dimensions 28x28 (MNIST), Target class

Output: Attacked Image

1:	**function** ALTERING MAXIMUM(Image,Target)
2:	Change the activation function of the Models last from softmax to linear
3:	Pick an image T from Target class
4:	Create a saliency map for both Image and T
5:	Round of values in the saliency maps to a single decimal place
6:	Find maximum occurring element M in the saliency map of Image
7:	Change the intensity to 0 in Image, for positions in Image with the saliency M
8:	Find maximum occurring element M1 in the saliency map of T
9:	Change the intensity to 175 in the Image, for positions in T with the saliency M1
10:	return Image
11:	**end function**

---

4   Results and Performance Analysis

Different layers and activation functions are used in CNN model training for MNIST and CIFAR-10 dataset. CNN for MNIST dataset has five set of convolutional layers includes the filter sizes as 64, 128, 256, 512 and 512 whereas, for CIFAR-10 dataset, it has two layers 32, 64 respectively. The images are convolved with the filters at each convolutional layer, followed by fully connected layers and a softmax layer. Rectified Linear Unit, ReLU is the activation function used at the convolutional layers [9]. The techniques applied on MNIST dataset to perform untargeted attacks are: (i) modification of background pixels, (ii) edge detection, (iii) combination of edge detection and background modification and, (iv) addition of noise and edge detection. Similarly for CIFAR-10 dataset, edge detection and combination of heat map generation with edge detection were applied. The size and accuracies obtained from both datasets are mentioned in Table 2.

Table 2. Summary of accuracy obtained for the CNN model trained over various datasets

Dataset	Training set		Validation set	
	Size	Accuracy (%)	Size	Accuracy (%)
MNIST	55000	99.82	5000	99.30
CIFAR-10	45000	97.23	5000	79.32

In Fig. 2, untargeted misclassifications caused by the techniques applied on MNIST dataset are shown. It shows the original image with its class against the adversarial image with the predicted class. The various untargeted attacks described were applied on the test set of MNIST and CIFAR-10 datasets, where the test sets contain 10000 images each. In Table 3, the summary of various techniques applied and its percentage of misclassification caused is tabulated. In addition, the performance of untargeted attacks is graphically represented in Fig. 3.

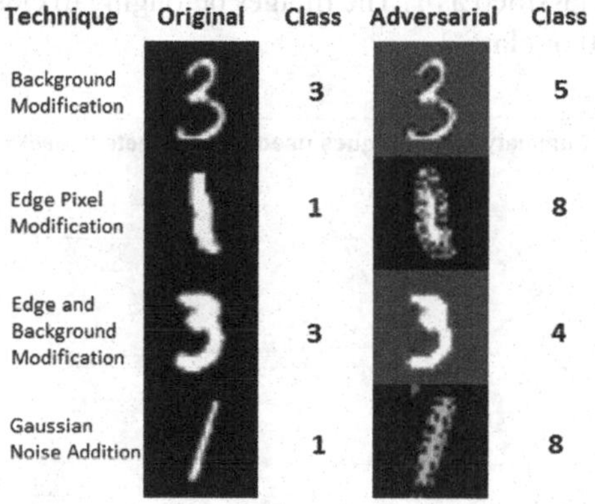

Technique	Original	Class	Adversarial	Class
Background Modification		3		5
Edge Pixel Modification		1		8
Edge and Background Modification		3		4
Gaussian Noise Addition		1		8

Fig. 2. Untargeted attacks – MNIST dataset

Table 3. Summary of untargeted attacks

Name of the technique	Dataset used	Misclassification (%)
Background modification	MNIST	68
Edge detection	MNIST	37
Combination of edge detection and Background modification	MNIST	89
Noise addition and edge detection	MNIST	32
Edge detection	CIFAR-10	47
Combination of heatmap and edge detection	CIFAR-10	43

In case of MNIST dataset, highest numbers of misclassifications were caused by the combination of Edge Detection and Background Modification. Though, Background Modification provides a good amount of misclassification, Edge Detection does not match it. This could be due to the difference in the number of pixels altered. There is also an added amount of uncertainty due to AES algorithm. Noise addition caused the least amount of misclassification proving that DNNs are not vulnerable to random changes in an image.

The results on CIFAR-10 shows that it is harder to generate robust adversarial images when compared to MNIST because of the three channels involved in colored images. Edge Detection fares better when compared to combination of Heatmap and Edge Detection.

Targeted attacks have been applied on a subset of the MNIST dataset. The model is trained on the images of numbers 0, 1, 2 and 7. The division of the dataset is shown in Table 4. Various techniques used are altering effective pixels using fixed values, altering effective pixels using AES and altering pixels with maximum occurrence. Figures 4 and 5, shows the targeted misclassifications caused on the subset of MNIST dataset, that is images with labels 0, 1, 2 and 7. For maximum occurring pixels, the images could not be modified to be misclassified as 1. The images belonging to class 3 (number 7) are not misclassified as class 0 or class 2.

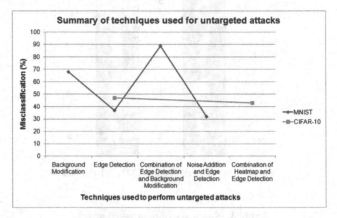

Fig. 3. Techniques used vs misclassification (%) - untargeted attacks

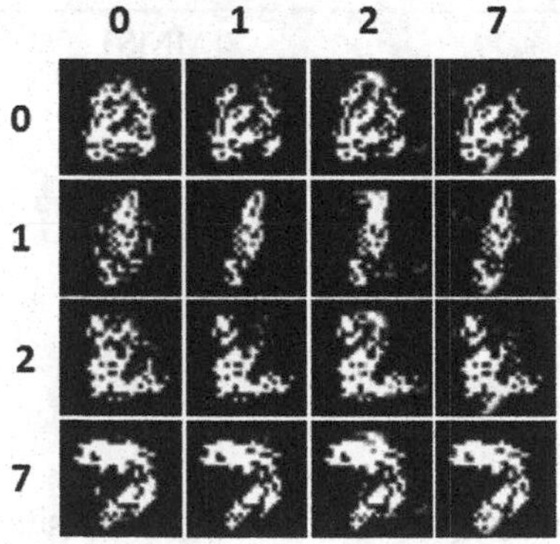

Fig. 4. Altering pixels with fixed values on MNIST dataset

In Table 5, summary of the targeted techniques applied and its percentage of misclassification caused are tabulated. Altering pixels with fixed values fares better than altering pixels with maximum occurrences in terms of misclassification because larger number of pixels are being modified. But the latter fares better in terms of visibility. The performance of the targeted attacks is shown in Fig. 6.

Various untargeted and targeted attacks were applied on datasets such as MNIST and CIFAR-10. On comparing the two datasets, MNIST dataset has black and white

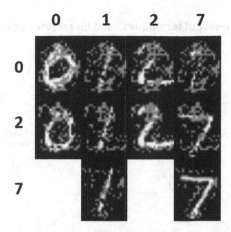

Fig. 5. Altering pixels with maximum occurrence on MNIST dataset

Table 4. Division of images in dataset

Class	Number of images in training set	Number of images in testing set
0	5923	980
1	6742	1135
2	5958	1032
7	6265	1028

Table 5. Summary of targeted attacks

Target	Misclassifications (%)	
	Altering pixels with fixed values	Altering pixels with maximum occurrences
0	49	32
1	77	0
2	96	71
7	71	5

images which uses single channel making it relatively easier to cause misclassification by making subtle changes when compared to CIFAR-10 having colored images with three channels. Since the untargeted attacks on CIFAR-10 did not yield results matching MNIST, targeted attacks were restricted to a subset of MNIST.

Untargeted attacks required the model to misclassify the image whereas targeted attacks required the model to misclassify the image to a particular class. The most successful untargeted attacks not only caused great amount of misclassifications but also made less prominent changes to the image. Comparatively, the targeted attacks required

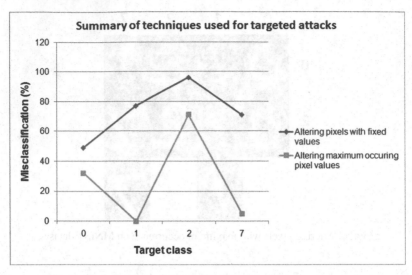

Fig. 6. Techniques used vs misclassification (%) - targeted attacks

more visible changes in order to misclassify to the target class. This research work is an attempt, to prove that Convolutional Neural Networks are vulnerable to adversarial changes being made in images of different datasets.

5 Conclusion

Adversarial attacks are a form of cyber attacks on machine learning models where the attacker inputs intentionally modified examples in such a way that the changes are undetectable to the human eye but causes the model to make a mistake during classification or prediction. There are two types of adversarial attacks, targeted and untargeted. While an untargeted attack aims to misguide the Convolutional Neural Network (CNN), a targeted attack aims to misguide the CNN to a specific class.

In this research work, different attacks based on pixels were created in order to gain an understanding on adversarial attacks. The pixels present in the background and/or on the edges where modified by adding a specific value or by encrypting them using AES algorithm. Other methods that were implemented includes, heatmap generation using Gradient-weighted Class Activation Mapping (GRAD-CAM) and noise addition with edge pixel modification. Targeted attacks involved generating saliency maps to alter pixels having a particular value or to alter pixels with the maximum occurrences. Finally, the effectiveness and shortcomings of each technique were inferred and analyzed.

The existence of adversarial attacks limits the areas in which deep learning can be applied, especially for security-critical tasks. This is mainly because of their good transferability, i.e., the adversarial attacks crafted for one model remains effective for others. As a future addition, defence mechanisms to prevent adversarial attacks can be implemented.

Acknowledgements. The authors sincerely thank Dept. of CSE, Sri Sivasubramaniya Nadar College of Engineering for the permission given to utilize the High Performance Computing Laboratory and its GPU server for successful execution of this project.

Conflict of Interest. The authors confirm that there is no conflict of interest for this publication.

References

Carlini, N., Wagner, D.: Towards evaluating the robustness of neural networks. In: IEEE Symposium on Security and Privacy (SP), pp. 39–57 (2017)

Chakraborty, A., Alam, M., Dey, V., Chattopadhyay, A., Mukhopadhyay, D.: Adversarial attacks and defences: a survey machine learning arXiv:1810.00069v1 (2018)

Chen, P.Y., Zhang, H., Sharma, Y., Yi, J., Hsieh, C.J.: ZOO: zeroth order optimization based black-box attacks to deep neural, networks without training substitute models. In: ACM Workshop on Artificial Intelligence and Security (AISec@CCS), pp. 15–26 (2017)

Dong, Y., et al.: Boosting adversarial attacks with momentum. In: IEEE/CVF Conference on Computer Vision and Pattern Recognition (CVPR), pp. 9185–9193 (2018)

Goodfellow, I.J., Shlens, J., Szegedy, C.: Explaining and harnessing adversarial examples. In: International Conference on Learning Representations, pp. 1–11 (2015)

Kurakin, A., Goodfellow, I., Bengio, S.: Adversarial machine learning at scale. In: International Conference on Learning Representations, pp. 1–17 (2015)

Su, J., Vargas, D.V., Sakurai, K.: One pixel attack for fooling deep neural networks. IEEE Trans. Evol. Comput., 1–12 (2019). https://doi.org/10.1109/TEVC.2019.2890858

Szegedy, C.: Intriguing properties of neural networks. In: International Conference on Learning Representations, pp. 1–10 (2013)

Yuan, X., He, P., Zhu, Q., Li, X.: Adversarial examples: attacks and defences for deep learning. IEEE Trans. Neural Netw. Learn. Syst., 1–20 (2019). https://doi.org/10.1109/TNNLS.2018.2886017

Zhao, Y.S., Zhu, H., Shen, Q., Liang, R., Chen, K., Zhang, S.: Practical adversarial attack against object detector. In: USENIX Security Symposium arXiv:1812.10217v3 (2018)

CIFAR–10, March 2020. https://www.cs.toronto.edu/~kriz/cifar.html

MNIST, March 2020. http://yann.lecun.com/exdb/mnist/

VGG-16 trained on CIFAR-10, April 2020. https://github.com/mjiansun/cifar10-vgg16

Ilyas, A., Santurkar, S., Tsipras, D., Engstrom, L., Tran, B., Madry, A.: Adversarial examples are not bugs, they are features. Computer Vision and Pattern Recognition (2019). arXiv:1905.02175v4

Vargas, A.V., Su, J.: Understanding the one-pixel attack: propagation maps and locality analysis (2019). arXiv:1902.02947v1

Continual Learning for Classification Problems: A Survey

Mochitha Vijayan[✉] and S. S. Sridhar

SRM Institute of Science and Technology, Chennai, India
{mv2820,sridhars}@srmist.edu.in

Abstract. Artificial Neural Networks performs a specific task much better than a human but fail at toddler level skills. Because this requires learning new things and transferring them to other contexts. So, the goal of general AI is to make the models continually learning as in humans. Thus, the concept of continual learning is inspired by lifelong learning in humans. However, continual learning is a challenge in the machine learning community since acquiring knowledge from data distributions that are non-stationary in general leads to catastrophic forgetting also known as catastrophic interference. For those state-of-art deep neural networks which learn from stationary data distributions, this would be a drawback. In this survey, we summarize different continual learning strategies used for classification problems which include: Regularization strategies, memory, structure, and Energy-based models.

Keywords: General AI · Artificial neural networks · Lifelong learning · Continual learning · Catastrophic forgetting · Classification problems

1 Introduction

Deep learning holds state-of-the-performances in specific tasks in the areas of Computer vision, Natural Language Processing (NLP), Speech recognition, etc. This success is due to supervised training from huge and fixed datasets. No matter how big your training dataset is, as we grow in terms of data dimensionality and when we would like to solve more complex problems (not the toy problems), it becomes exponentially more difficult to cover the entire space of possibilities without representing the data set collected a priori. But how can we improve the scalability of models in terms of computation and memory resources and how can they adapt to new circumstances never seen before? The answer to this is to never stop learning as our brain does [7, 8]. So, Continual Learning [1, 2] which is inspired by lifelong learning in humans is the way to deal with a higher and realistic time-scale where data (and tasks) become available only during the time.

Why continual Learning is a challenging problem?

Many connectionist models that are gradient-based such as Artificial Neural Network's (ANN) suffer from Catastrophic forgetting [3, 4]. This means that once you train a network on a piece of data and when we expose it to a little new data distribution,

© IFIP International Federation for Information Processing 2021
Published by Springer Nature Switzerland AG 2021
V. Krishnamurthy et al. (Eds.): ICCIDS 2021, IFIP AICT 611, pp. 156–166, 2021.
https://doi.org/10.1007/978-3-030-92600-7_15

it tends to forget the older ones. Since every parameter of the network is rewritten to suit the new data distribution. ANNs are unable to learn new information or instances immediately. Deep NNs are the dominant approach to machine perception, but: 1. They cannot learn new instances immediately, 2. Learning requires multiples loops over a data set, 3. They are susceptible to catastrophic forgetting if the data is not iid (independently and identically distributed).

To overcome catastrophic forgetting AI systems must be able to retain the previously learned task, acquire new knowledge and restrict the novel data from interfering with the existing knowledge. That is, the system must be stable to retain acquired knowledge without catastrophically forgetting them and must be plastic enough to integrate novel information. This concept is being widely explored in both biological and artificial models and is well-known as the stability-plasticity dilemma [5, 6]. Two reasons to solve catastrophic forgetting: 1. Making systems learn like humans and animals [35, 37], 2. Enabling new applications like immediate inference from new input labels, to avoid retraining from scratch thereby optimizing memory and computational resources.

For a classification task, Task-IL is often artificial, Class-IL is crucial for a continual learning setting in classification problems. This survey focuses on different continual learning strategies used for classification tasks. Furthermore, some ideas to enhance continual learning efficiency is also being discussed. We survey various strategies for continual learning in neural networks that mitigate catastrophic forgetting to different levels. We categorize the continual learning approaches used in the existing systems into Regularization strategies, Memory-based strategies, Structure-based strategies, and Energy-Based Models.

2 Regularization Strategies

In continual learning, regularization is used to ensure the stability of important parameters for the previously learned tasks.

Li et al., 2018 present Learning without Forgetting (LwF) [12] which is a hybrid of knowledge distillation (distillation loss to maintain consistency) and fine-tuning. This method has a relation with replay-based methods in a way that rather than storing or generating the samples to be replayed. Using the model learned on the previous tasks this approach labels the current task and replays them. Trying to transfer the information from an extremely regularized massive model to a smaller model. For a parameter set θ_s (shared) of all the tasks it tries to optimize the parameter θ_n of the novel task along with θ_s. An extra restraint is imposed such that for a new task the parameters θ_s and the parameters of old tasks θ_o will not deviate much to remember θ_o. If for a new task, (X_n, Y_n) is the training data, the old tasks output Y_o, and new parameters θ_n(randomly choosed), then the updated parameters $\theta_s^*, \theta_o^*, \theta_n^*$ are:

$$\theta_s^*, \theta_o^*, \theta_n^* \leftarrow argmin_{\hat{\theta}_s, \hat{\theta}_o, \hat{\theta}_n} \left(\lambda_o \mathcal{L}_{old}\left(Y_o, \hat{Y}_o\right) + \mathcal{L}_{new}\left(Y_n, \hat{Y}_n\right) + R\left(\hat{\theta}_s, \hat{\theta}_o, \hat{\theta}_n\right) \right)$$

Here a loss is added to discourage the old task *output to change* for new tasks. Optimization is done both for the final layers and shared representation. They show that LwF performs better than LFL [13] which is a similar approach. The drawback of this approach is that it extremely depends on the task's relevance and the training time for one task will have a linear hike with the number of tasks learned. In multi-task learning distillation used here is a potential solution but it demands a repository of persistent data for every task learned. This method is not immediately applicable in Reinforcement Learning scenarios.

In Less-forgetting learning (LFL), [13], the old task performance is preserved by protecting the shared representations. This learning model satisfies two properties, there should be clear-cut boundaries for each task and the shared parameters should be kept unchanged. For the final most hidden activations the l2 distance between them is regularized, thereby protecting the previous input to output association being learned from the old tasks. This is done by the computation of additional activations using the parameters of old tasks. This L2 loss discourages the *output after* θ_s (shared parameter set) from changing for new tasks, while θ_o (task-specific parameters for past tasks) remains as it is. However, it requires the computation of the old task parameters for each new data point making it computationally expensive.

Kirkpatrick et al., 2017 developed an algorithm akin to synaptic consolidation for Artificial Neural Nets known as EWC (Elastic Weight Consolidation) [10] which focuses on task-specific synaptic consolidation. EWC allows the knowledge acquired in the previous task to be protected during a new task learning as shown in Fig. 1.

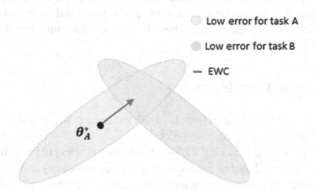

Fig. 1. EWC assures task A is retained while training on task B.

The importance of the parameter θ concerning a task's training data D is represented as $p(\theta \mid D)$, the posterior distribution. While considering a scenario with tasks A (with D_A) and B (with D_B) which are independent, applying Baye's theorem to the posterior probability and then taking the log gives us,

$$\log p(\theta|D) = \log p(D_B|\theta) + \log p(\theta|D_A) - \log p(D_B)$$

The knowledge of earler task is held by $\log p(\theta \mid D_A)$, the posterior probability. EWC uses a Gaussian distribution and adds a cost term to prioritize the important weights of past tasks by using a fisher information matrix F and is evaluated by its diagonal.

Gaussian distribution is used since $\log p\,(\theta \mid D_A)$ is intractable.

$$\mathcal{L}(\theta) = \mathcal{L}_B(\theta) + \sum_i \frac{\lambda}{2} F_i \left(\theta_i - \theta_{A,i}^*\right)^2$$

EWC performed well in both supervised (permuted MNIST task framed by [32]) and reinforcement (Atari 2600) learning scenarios. EWC's characteristic of making weights less plastic with time promotes memory retention rather than forgetting. As the number of random patterns exceeds the capacity of the network EWC degrades than the plain gradient descent. Moreover, Hopfield networks which is an EWC model suffer from the phenomenon of blackout catastrophe when the network capacity is saturated which results in the hindrance of retrieval of old knowledge and addition of new ones. The available output labels have to be summed up to calculate the diagonal of the Fisher matrix which makes its complexity linear to the number of outputs, which limits the application of this algorithm to low-dimensional output spaces. [34] show that EWC works poorly at incremental class learning.

Zenke et al., 2017 introduced intelligent synapses [11] by maintaining an online measure of synapses (parameters like weights and bias) "importance" in solving the past tasks. These important synapses are protected from changing when the task changes. That is, the future learning is assisted by the synapses which are of less significance to the previous tasks, thereby alleviating catastrophic forgetting. A surrogate loss term is added to hinder significant variations in important parameters θ_k while learning a novel task.

Loss function:

$$\mathcal{L}'_\mu = \mathcal{L}_\mu + c \underbrace{\sum_k \Omega_k^\mu \left(\theta_k' - \theta_k\right)^2}_{\text{Surrogate loss}}$$

This measure tracks parameters (past and present) and evaluates their importance online. Parameters contributing more to the loss function are more relevant. In [11], SI is shown to perform well on a multi-headed Split MNIST of their design and had shown similar performance as EWC on the permuted MNIST dataset. By using a multi-headed split CIFAR task of their design some transfer learning is also done.

In [28] (Lee et al., 2017), A L2 penality is applied to the shared parameter changes and for learning a novel task. Models formed are merged using first or second-moment matching. IMM is a progression of EWC which performs a model-fusion separately upon learning a novel task.

Aljundi et al., 2018 proposed Memory-aware Synapses (MAS) [14] which is a model-based method inspired by neuroplasticity and Hebbian learning in biological systems. MAS determines the significance of the parameters that adapt to the test data set using the variation in the model output function concerning the inputs, rather than the loss function. Thus, avoiding the problem of falling in local minima. Important parts of the model can be learned using unlabeled data. This computation is done online and in an unsupervised manner. MAS could achieve in terms of constant memory w.r.t the task count, ability to build on the top of a pre-trained model and to add new tasks, ability to learn from unlabelled data.

Uncertainty-guided Continual learning approach with Bayesian neural networks (UCB) [15] (Ebrahimi et al., 2020), used an uncertainty prediction strategy for continual learning. Parameters of importance are either fully safeguarded using a binary mask (UCB-P) or can be changed conditionally based on their uncertainty learning novel tasks (UCB). UCB-P avert forgetting after the initial phase of pruning by retaining a binary mask for each task. UCB doesn't take additional memory and let on a more flexible learning mechanism in the network by limiting the forgetting.

Although these regularization-based approaches can be computationally efficient in alleviating catastrophic forgetting under certain conditions, a disadvantage is that they gradually reduce the model's capacity for learning new tasks. In class-incremental learning, regularization-based methods are shown to consistently fail [30, 31].

3   Memory-Based Strategies

A set of experiences, exemplars, or vectors which represent the tasks can be stored rather than storing all the observations. This provides a more efficient and scalable memory strategy. This also enables compression and high-level transfer across multiple tasks. The focus of many works is dealing with the challenge to determine which samples to store.

Robins, A., 1995 in his work [25] interleaved new experiences with the generated patterns of previous experiences intending to make neural networks able to encode information, store, and recall them on the need for better generalization across tasks. This method performed well in alleviating catastrophic forgetting. i.e., a continual rehearsal of previous tasks is done. This continual learning strategy based on the memory is called replay or otherwise rehearsal.

Rebuffi et al., 2017 proposed iCaRL (Incremental Class and Representation Learning) [9] which combines regularization and memory strategies. iCaRL simultaneously learns the classifier and feature extractor. It consists of three components: classification done by a nearest-mean exemplars rule, herding based exemplar selection, learning of representation that uses exemplars along with distillation intending to avoid catastrophic forgetting. One exemplar image each for all classes seen so far is stored. Then computes a model for each of the classes which is the average of each exemplar stored. The input image is classified as the class of the prototype that is close to the current feature representation. As the feature representation changes recompute the mean of exemplars.

The training set consists of samples of new data as well as the exemplars of old classes (to remind what the old classes were).

$$D \leftarrow \bigcup_{y=s,..,t} \{(x,y) : x \in X^y\} \cup \bigcup_{y=1,..,s-1} \{(x,y) : x \in P^y\}$$

Loss function:

$$\mathcal{L}(\theta) = - \sum_{(x_i,y_i)\in D} [\underbrace{\sum_{y=s}^{t} \delta_{y=y_i} \log(g_y(x_i))}_{\text{classification loss}} + \underbrace{\sum_{y=1}^{s-1} q_i^y \log(g_y(x_i))}_{\text{distillation loss}}]$$

The classification loss term will help to learn new classes and the distillation loss term will help not to forget previous outputs.

iCaRL can learn in an incremental fashion over a long period in experiments conducted on CIFAR-100 and ImageNet ILSVRC 2012 data wherein other methods fail quickly. Whereas, iCarL's performance is not appreciable when the training samples of classes are available in batch settings. iCarl's performance is heavily influenced by the number of examples it stores.

Variational Continual Learning (VCL) by Nguyen et al., 2018 [27] uses Variational Interference (VI) whic is also a replay-based method. The previous posteriori is multiplied by the likelihood of the new task data set to obtain the new posterior distribution. This is a Bayesian inference to carry out continual learning. They calculate the Kullback-Leibler (KL) (approximately a quadratic regularization with rotation) divergence between the current distribution and the previous posterior. In addition to the VCL regularization term they showed that by using a core-set, which samples examples from old tasks, VCL is experiencing less forgetting. VCL with or without core-set outperforms EWC and SI on the permuted MNIST and multi-headed split MNIST dataset. Using permuted MNIST shows to yield misleading results in the context of continual learning as shown in [29].

A replay is a phenomenon seen in rodents and humans [24]. But it is unrealistic to maintain an unlimited number of observations or exemplars. In generative models as in [23], samples are not stored. Instead, generative models are trained and then used to produce data needed for rehearsal. Trying to mimic the generative nature of the hippocampus for the rehearsal of past experiences, Shin et al., 2017 proposed Deep Generative Replay (DGR) [23] consisting of two components: a generative model and a solver. A pseudo-data that represents the previous tasks are generated and are interleaved along with new tasks. This avoids the need for explicitly revising past training data for rehearsal, thus downsizing memory requirements. The catastrophic interference problem now shifts to the training process of the generative model.

In [18] (Kemker et al., 2018) proposed a model for Class-IL inspired by the dual memory model [3] of a mammalian brain and the studies of recall and consolidation that happens in a mammalian brain while fear conditioning [36]. FearNet uses a hippocampal net that is responsible for recalling immediate memories, long-term (mature) memories are handled by a PFC network and a neural net that took inspiration from the basolateral amygdala for deciding whether the model should make use of the hippocampal or PFC network for that specific sample or not. Memory consolidation happens during the sleeping phase that the FearNet consolidates hippocampal network to the PFC network, hippocampal network and PFC operates as complementary memory systems. PFC is a generative neural net that generates pseudo examples that are interleaved with the new samples in the hippocampal network.

One way to enforce the gradient to stay close to the gradients from previously learned tasks is to eliminate its interference as in [19, 20]. These methods are beneficial in multi-task learning as they can make learning more efficient in case of opposing objectives. Lopez-Paz et al., 2017 introduced the Gradient Episodic Memory (GEM) [19] model that brings in the positive backward transfer to previously learned tasks. It is the first approach to accomplish learning in one pass. GEM is characterized by an episodic memory that stores a subset of the observed examples for a given task and alleviates catastrophic forgetting (negative backward transfer). This model enables to learn the subset of correlativity of a set of tasks, capable to predict desired values associated with past or novel tasks without enforcing task descriptors. GEM's memory requirement is high than other regularization strategies such as EWC [10] at the training time. As GEM is being evaluated on MNIST and CIFAR datasets it is doubtful if it scales to more realistic problems. An inherent problem here is the constraint on the amount of experience that can be stored in memory, which could quickly become a limiting factor in large scale problems. Even when this method works well in a single pass context it demands much more computation.

Chaudhry, A. et al., 2018 proposed Averaged GEM (A-GEM) [20] which is an advanced version of GEM which alleviates the computational burden of GEM. GEM ensures that the loss of each specific previous tasks (sampled from episodic memory) at every training step does not increase. Whereas, A-GEM ensures that the average of the episodic memory loss over the past tasks at every training step does not increase. A-GEM shows the same or even better performance as GEM. A-GEM is shown to be computationally and memory-efficient as EWC and other regularization-based methods.

Scalability of the approaches that use memory for replay or rehearsals when the number of task increases is arguable. And the rehearsal-based approaches have to deal with the quality of samples generated which would again be a drawback.

4   Structure-Based Strategies

These strategies determine if the network has to be expanded to represent the new tasks or not. Progressive networks [22, 26] are dynamic architecture-based approaches where the base architecture was replicated and some connections were included in response to new tasks. These approaches were successful in reinforcement learning scenarios.

Mallya et al., 2018 proposed PackNet [16]. Forgetting is alleviated by iterative pruning to completely freeze the updates on the most important parameters. A binary mask (task-specific) that indicates the importance and unimportance of parameters is saved. More specifically, a single network's capacity is used to learn multiple tasks and is done by freeing the parameters irrelevant to the present task depending on their magnitude. This approach requires knowing the task prior to the use of a suitable mask. Ranking the importance of weight by their magnitude will not be a guaranteed "importance" indicator. Parameter importance is ranked by its magnitude in PackNet which cannot give any guarantee to be an "importance" indicator. This model relies on the explicit parametrization of importance.

Hard attention to the tasks (HAT) [17] introduced by Serra et al., 2018 learns an attention vector to determine the important neurons (using mini-batch stochastic gradient descent and backpropagation) to the task controlling the gradient shifts. An almost-binary mask per previous tasks is used to maintain the information learned on previous tasks. HAT employs an attention-weighted L1 regularization on the attention mask Instead of simple L1 regularization. These attention masks are lightweight structures included without changing much in the existing network. When EWC and SI adds a soft architectural regularization to the loss function, HAT adds a hard structural regularization to both the loss function and gradient magnitudes.

Rusu et al., 2016 in their work [22] hindered any adjustments to the previous network. Instead expanded the network with a new subnetwork to be trained with new data. This progressive network retains a repository of pre-trained architectures for each learned task. If the features being learned cannot correctly represent the novel task, more neurons are added to the network to account for the features of the new task. This approach has shown favorable results for a wide variety of reinforcement learning problems when compared to other strategies that either priorly train or tweak the model incrementally with the prior knowledge only at the initialization phase. This strategy prevents catastrophic forgetting, however, without considering the difficulty of the task they add in a constant count of units for each learned task leading the complexity of the network architecture to grow and thus is suboptimal in terms of network capacity utility and performance.

Yoon et al., 2018 proposed Dynamically Expandable Networks (DEN) [21] in which, the network expands according to the task at hand by dividing/replicating the most important neurons retraining them on new tasks. This separate stage is known as selective retraining. The important drifting units (neurons) are identified using a complex mixture of hyperparameters and heuristics. They employ L1 regularization as well as L2-transfer to condition learning, regularization constants, and a set of extra thresholds. Even though they bring in high computation this is unavoidable in a continual learning environment where several tasks have to be learned and when network capacity cannot be fixed.

When new tasks are encountered modular-based models go for explicit ways to increase architectural capacity. In recent times, several adaptive network models have emerged that expand and at the same time protects and reuse the existing representations [22]. But this leads to demanding high computational requirements in the learning process. In [26] (Schwarz et al., 2018) increase networks capacity more aggressively if needed and then prune or sometimes compress the parts of the network.

Structure-based approaches are limited to task-incremental learning settings.

5   Energy-Based Models for Continual Learning

Li et al., 2020 proposed an Energy-based model [33] for continual learning which does not use an extra loss, memory, or model and is still performing well in class incremental learning scenarios and boundary agnostic settings which are the challenges of Continual Learning. EBMs naturally deals with the challenging problems in Continual Learning without replay. A contrastive divergence training procedure is used and it provides a natural way to deal with dynamically growing. They propose to learn new conditional gains during the training process, which makes EBMs parameter update cause less interference with old data. But EBM's takes more time than softmax-based classifier model in terms of convergence and evaluation.

6   Research Insights

- The model (agent) should be capable of learning from both stationary and non-stationary data streams. In real-world data comes from a dynamic data distribution that keeps on changing with time. The state-of-art neural networks are provided with iid data and are allowed to loop over it again and again.

- Data streams like MNIST and permuted MNIST are highly unrealistic, most of the conventional neural nets use them. Evaluation of the existing continual learning is mostly done on toy data streams. The applicability of these approaches in a real-world setting is questionable.
- Models that work well in MNIST, CIFAR, etc. tend to fail in other important paradigms as shown [34]. MNIST and CIFAR are small datasets.
- Memory and computations must be kept fixed across the models to enable a fair comparison which is not done in most of the cases.
- Evaluation of a model in terms of Task-IL, Domain-IL, and Class-IL can be done only on continual learning problems with a series of tasks with clear cut boundaries [31].

7   Concluding Remarks and Future Direction

We can make artificial agents perform a particular task surprisingly much better than a human. But current artificial agents are still in their infancy due to a serious problem known as catastrophic forgetting. Continual learning as in humans is inevitable. A continual learning agent must not suffer from the phenomena of catastrophic forgetting. It means that the agent must retain its ability to perform fairly well on previously learned tasks and should be able to learn new tasks by extracting knowledge from the previous ones exhibiting positive forward transfer thereby achieving better performance and fast learning. It should be scalable by dynamically adapting to the real-time environment with varieties of tasks. It should enable positive backward transfer by gaining better performance on previous tasks after learning a new similar task. The data collected from the real-world may not be always explicitly labeled. The agent should learn from unlabeled data and even should be the recipient of tasks without clear task boundaries. Catastrophic forgetting isn't the only barrier of general AI. Not every task has a huge dataset and the challenge is to make the models learn from more than just from the samples. This survey tries not only to highlight the inevitability of continual learning but also to report the limitations of existing state-of-art neural networks in this regard.

References

1. Hassabis, D., Kumaran, D., Summerfield, C., Botvinick, M.: Neuroscience-inspired artificial intelligence. Neuron Rev. **95**(2), 245–258 (2017)
2. Thrun, S., Mitchell, T.: Lifelong robot learning. Robot. Auton. Syst. **15**, 25–46 (1995)
3. McClelland, J.L., McNaughton, B.L., O'Reilly, R.C.: Why there are complementary learning systems in the hippocampus and neocortex: insights from the successes and failures of connectionist models of learning and memory. Psychol. Rev. **102**, 419–457 (1995)
4. McCloskey, M., Cohen, N.J.: Catastrophic interference in connectionist networks: the sequential learning problem. Psychol. Learn. Motiv. **24**, 104–169 (1989)
5. Ditzler, G., Roveri, M., Alippi, C., Polikar, R.: Learning in nonstationary environments: a survey. IEEE Comput. Intell. Mag. **10**(4), 12–25 (2015)

6. Mermillod, M., Bugaiska, A., Bonin, P.: The stability-plasticity dilemma: investigating the continuum from catastrophic forgetting to age-limited learning effects. Front. Psychol. **4**, 504 (2013)
7. Grossberg, S.: How does a brain build a cognitive code? Psychol. Rev. **87**, 1–51 (1980)
8. Grossberg, S.: Adaptive resonance theory: how a brain learns to consciously attend, learn, and recognize a changing world. Neural Netw. **37**, 1–41 (2012)
9. Rebuffi, S.-A., Kolesnikov, A., Sperl, G., Lampert, C.H.: iCaRL: incremental classifier and representation learning. In: Conference on Computer Vision and Pattern Recognition, Honolulu, pp. 5533–5542. IEEE (2017)
10. Kirkpatrick, J., et al.: Overcoming catastrophic forgetting in neural networks. Proc. Natl. Acad. Sci. USA **114**, 3521–3526 (2017)
11. Zenke, F., Poole, B., Ganguli, S.: Continual learning through synaptic intelligence. In: International Conference on Machine Learning, Sydney, pp. 3987–3995. PMLR (2017)
12. Li, Z., Hoiem, D.: Learning without forgetting. IEEE Trans. Pattern Anal. Mach. Intell. **40**, 2935–2947 (2017)
13. Jung, H., Ju, J., Jung, M., Kim, J.: Less-forgetting learning in deep neural networks. In: AAAI 2018, New Orleans, LA (2018)
14. Aljundi, R., Babiloni, F., Elhoseiny, M., Rohrbach, M., Tuytelaars, T.: Memory aware synapses: learning what (not) to forget. In: Ferrari, V., Hebert, M., Sminchisescu, C., Weiss, Y. (eds.) ECCV 2018. LNCS, vol. 11207, pp. 144–161. Springer, Cham (2018). https://doi.org/10.1007/978-3-030-01219-9_9
15. Ebrahimi, S., Elhoseiny, M., Darrell, T., Rohrbach, M.: Uncertainty-guided continual learning with Bayesian neural networks. In: ICLR 2020 (2020)
16. Mallya, A., Lazebnik, S.: PackNet: adding multiple tasks to a single network by iterative pruning. In: IEEE Conference on Computer Vision and Pattern Recognition (CVPR) (2018)
17. Serra, J., Suris, D., Miron, M., Karatzoglou, A.: Overcoming catastrophic forgetting with hard attention to the task. In: Dy, J., Krause, A. (eds.) Proceedings of the 35th International Conference on Machine Learning. Volume 80 of Proceedings of Machine Learning Research, pp. 4548–4557. PMLR (2018)
18. Kemker, R., Kanan, C.: FearNet: brain-inspired model for incremental learning. In: ICLR 2018 (2018)
19. Lopez-Paz, D., et al.: Gradient episodic memory for continual learning. In: Advances in Neural Information Processing Systems (NeurIPS), p. 30 (2017)
20. Chaudhry, A., et al.: Efficient lifelong learning with A-GEM. arXiv https://arxiv.org/abs/1812.00420, 2 December 2018
21. Yoon, J., Yang, E., Lee, J., Hwang, S.J.: Lifelong learning with dynamically expandable networks. In: International Conference on Learning Representations (2018)
22. Rusu, A., et al.: Progressive neural networks. arXiv preprint arXiv:1606.04671 (2016)
23. Shin, H., et al.: Continual learning with deep generative replay. In: Advances in Neural Information Processing Systems (2017)
24. Liu, Y., et al.: Human replay spontaneously reorganizes experience. Cell **178**, 640–652 (2019)
25. Robins, A.: Catastrophic forgetting, rehearsal and pseudorehearsal. Connect. Sci. **7**, 123–146 (1995)
26. Schwarz, J., et al.: Progress & compress: a scalable framework for continual learning. In: Proceedings of the International Conference on Machine Learning, pp. 4535–45442 (2018)
27. Nguyen, C.V., Li, Y., Bui, T.D., Turner, R.E.: Variational continual learning. In: International Conference on Learning Representations (2018)
28. Lee, S.-W., Kim, J.-H., Jun, J., Ha, J.-W., Zhang, B.-T.: Overcoming catastrophic forgetting by incremental moment matching. In: Advances in Neural Information Processing Systems, pp. 4652–4662 (2017)

29. Lee, S.W., Kim, J.H., Ha, J.W., Zhang, B.T.: Overcoming catastrophic forgetting by incremental moment matching. arXiv preprint arXiv:1703.08475 (2017)
30. Farquhar, S., Gal, Y.: Towards robust evaluations of continual learning. arXiv preprint arXiv: 1805.09733 (2018)
31. van de Ven,G.M., Tolias, A.S.: Three scenarios for continual learning. arXiv preprint arXiv: 1904.07734 (2019)
32. Goodfellow, I.J., Mirza, M., Xiao, D., Courville, A., Bengio, Y.: An empirical investigation of catastrophic forgetting in gradient-based neural networks. arXiv 4.2.1, A.1, A.5 (2013)
33. Li, S., Du, Y., van de Ven, G.M.: Energy-based models for continual learning. arXiv preprint (2020)
34. Kemker, R., McClure, M., Abitino, A., Hayes,T.: Measuring catastrophic forgetting in neural networks. In: Proceedings of the AAAI (2018)
35. Hadsell, R., Rao, D., Rusu, A.A., Pascanu, R.: Embracing change: continual learning in deep neural networks. Trends Cogn. Sci. 24(12), 1028–1040 (2020)
36. Kitamura, T., Ogawa, S.K., Roy, D.S., Okuyama, T., Morrissey, M.D., Smith, L.M., et al.: Engrams and circuits crucial for systems consolidation of a memory. Science 356, 73–78 (2017)
37. Parisi, G.I., Kemker, R., Part, J.L., Kanan, C., Wermtera, S.: Continual lifelong learning with neural networks: a review. Neural Netw. 113, 54–71 (2019)

Detection of Flooded Regions from Satellite Images Using Modified UNET

S. M. Jaisakthi[1], P. R. Dhanya[1]([⊠]), and S. Jitesh Kumar[2]

[1] School of Computer Science and Engineering, Vellore Institute of Technology, Vellore, India
{jaisakthi.murugaiyan,dhanya.pr}@vit.ac.in
[2] Sri Venkateshwara College of Engineering, Chennai, India
2018cse0716@svce.ac.in

Abstract. It is necessary to analyze the accessibility of flood-affected regions for better planning of response. Flood level detection using remotely sensed images could minimize costs and can also allow taking adequate preparation for fast recovery. MediaEval 2017 multimedia satellite task dataset is used for flood detection in satellite images. The image segmentation technique on satellite images using the modified UNET convolutional neural network is applied after pre-processing to analyze flood-affected regions and compared them to the corresponding segmented regions during the period of a flood event. Initially, all images were segmented using conventional segmentation methods like the Mean shift clustering algorithm. It is observed that the U-Net model produced a good measure for Intersection over Union (IoU) as 99.46% with 99.41% accuracy. The segmented images are then considered for further conventional processing to get meaningful information of flooded regions from satellite imagery.

Keywords: Flooded region segmentation · Convolutional neural network (CNN) · UNET

1 Introduction

A flood event is a major disaster that causes several damages and needs emergency responses. The information about road conditions and accessibility is essential for fast response and recovery during flooding. The inundated roads are often the leading cause of harm to lives from most calamities. Newly discovered tools using satellite imagery could reduce monitoring costs and it also helps to take necessary and immediate response to floods. Satellite images should be pre-processed because they may contain noise such as vegetation, varying illumination conditions, and atmospheric distortion. Flooded regions in the satellite images can be segmented using Deep Neural Networks. The convolution Neural

Supported by Department of Science and Technology, Science and Engineering Research Board, Government of India (DST-SERB).

© IFIP International Federation for Information Processing 2021
Published by Springer Nature Switzerland AG 2021

V. Krishnamurthy et al. (Eds.): ICCIDS 2021, IFIP AICT 611, pp. 167–174, 2021.
https://doi.org/10.1007/978-3-030-92600-7_16

Networks algorithm makes use of previous and current imagery of flood events. The CNN algorithm detects irregularities from a pre-flood stage by analysing the difference between the flooded and non-flooded images of a region with segments.

The Convolutional, Relational, and Dilated convolutional networks were used to segment flooded regions from satellite images with 88.23% of Mean Intersection over Union [1]. A convolutional InceptionV3 network using the pre-trained weights on ImageNet [2] and finely refine the last inception model in the network used for the segmentation task [3]. A fully convolutional network was also used for semantic segmentation of flooded regions and obtained good results when incorporating RGB and Infrared channels into the model. It shows IR information has vital importance for detecting flooded areas in satellite imagery [4]. Accuracy and the prediction performances of recent deep learning algorithms such as SegNet, UNet, PSPnet, and FCN were compared and analysed for the water segmentation task [5]. A Generative Adversarial network is also used for the segmentation task with an IoU of 0.83 [6]. Deep convolutional neural networks pre-trained on ImageNet along with a transfer learning mechanism, is employed to detect if an area has been affected by a flood in terms of access. The best models achieved averaged F1-Score value of 73.27% on the segmentation task [7]. The conventional methods employed to study and compare various methods for detecting and road passability by analysing the tweets which share images of flood and its metadata [8]. A Deep neural network ResNet50 pre-trained on ImageNet and Support Vector Machine used for the flood detection task with Twitter data [9]. Some studies depend on a convolutional neural network (CNNs) and a transfer learning-based classification method attains the F1-score of 62.30% and 61.02% for two runs [10]. A methodology adopted was based on a threshold decision tree classification of a simple three-band difference image, incorporating a newly developed "woody index", ETM+ Band 2 and ETM+ Band 5 [11]. A Difference of Normalized Difference Water Indices (DNDWI) that maps flooded regions received an accuracy of 85.68% and the non-flooded regions received an accuracy of 92.13% [12]. A reliable and flexible system for the contextual and semantic embellishment of satellite imagery by analyzing multimedia content from the social media platform uses Twitter as the prime data source for collecting data and computing multimedia content such as images [13]. The mask R-CNN and Fully Convolutional Network (FCN) are compared to select the apt model for the image segmentation task [14]. A study concentrated on the various techniques derived from the two case studies for flood extent mapping using the fusion of the MODIS imagery and passive microwave [15]. A paper provides a summary of remote sensing and modelling practices for predicting the vulnerability to flooding, calculating the intensity of flooding, and analysing the destruction caused by the flood [16]. A study demonstrates data taken from non-authoritative sources such as Volunteered Geographic Information, tweets, real-time traffic conditions and a list of bridge closures and Calgary road are used to extend traditional data and methods for flood extent mapping to identify flood-affected roads [17]. The Experimental results of a study show that integrating local semantic information yields slightly better performance than only using image-level CNN representation. Text features are not competitive when considering social media images [18]. It observed that media persons are required to contact by analysing different communication channels such as social media in particular, during flood events [19].

In this paper, a modified U-Net is employed to segment flooded regions in satellite images. The paper uses the Mediaeval 2017 dataset for the processing and evaluation of proposed model. A pre-trained MobileNetV2 used as an encoder and the decoder is the upsample block. The modified UNET model gives good values for accuracy and Intersection over Union (IOU). Different types of optimizers, loss functions and activation functions tried with the proposed model. The best functions selected for compiling and training the model.

2   Proposed Method

2.1   Dataset

This paper is intended to develop a method using deep neural networks that can recognize flood-affected regions in satellite imagery. This paper uses the MediaEval 2017 multimedia satellite task dataset for the segmentation task. The dataset comprises satellite image patches obtained from Planet's 4-band satellites with the shape of $320 \times 320 \times 4$ pixels and patches are in the form of GeoTiff format. Each image patch gives four-channel information such as with Red, Green, Blue, and Near-Infrared band. The dataset included 462 image patches from six different locations, and also its pixel values are in the 16-bit digital number format. They provided a segmentation mask of the flooded area and extracted it by human annotators (0 represents background and 1 for the flooded area) for each patch [20].

2.2   Methodology

Initially, the dataset is processed with a data augmentation technique which is used to improve the diversity of the training set by adding some random but realistic transformations. Augmentation techniques such as random rotation and horizontal flipping are used to increase diversity. There are many traditional methods for the segmentation of images such as point, line, and edge detection methods, morphological approaches, thresholding, region-based and pixel-based clustering. Different kind of methods developed for segmentation using convolutional neural networks that have become the most relevant technology to handle advanced challenges in the area of image segmentation. A modified U-Net model [21] used to segment flooded regions in satellite images (see Fig. 1). The satellite image is reshaped to 224×224. A typical U-Net model comprises an encoder and a decoder. A pre-trained model is used as the encoder to learn significant features and lower the number of trainable variables. Thus, the encoder for the segmentation task is a pre-trained MobileNetV2 [22] model and its intermediate outputs are used. The decoder for the model is the upsample block.

Different types of optimizers and losses are tried to compile the model. An adam optimizer, replacement optimization algorithm for stochastic gradient descent with a learning rate of 1e−3 has used when compiling the model.

The model has compiled with a loss of Mean Squared Error (MSE). It computes the mean of squares of errors between labels and predictions. The model trained with 500 epochs and a batch size of 64. After training the model, an Otsu thresholding [23] algorithm that returns a single intensity threshold used to separate pixels into two classes, foreground and background.

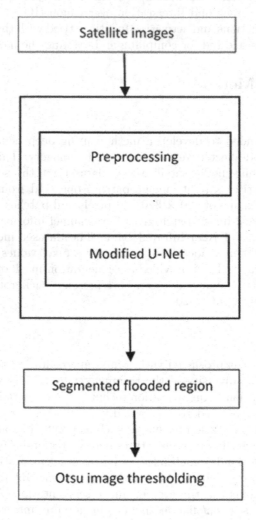

Fig. 1. Proposed method.

3   Result and Discussion

Intersection over Union (IoU) metric i.e. the Jaccard Index used for the primary and formal evaluation that is IoU = True Positive/(True Positive + False Positive + False Negative). It determined over the test dataset to analyse the efficiency of generated segmentation masks for flooded regions in the satellite image patches. The IoU measures [24] the accuracy for the pixel-wise classification. The dice coefficient is given by the ratio of, two times the Area of Overlap and the total number of pixels in both images. Soft Dice loss is given by the difference of Dice coefficient from unity. Initially, satellite images were processed with a mean shift clustering algorithm. But when analyzing the values for segmenting, it is clear that it does not give better value for IoU. So metrics for the proposed modified U-Net model has given in Table 1.

Table 1. IoU, soft dice loss, loss and accuracy of each dataset

Splitting of dataset	IoU (%)	Soft dice loss (%)	Loss (%)	Accuracy (%)
Training set	99.46	1.64	0.46	99.41
Validation set	71.35	49.59	27.57	68.55
Test set	71.55	47.71	27.72	68.78

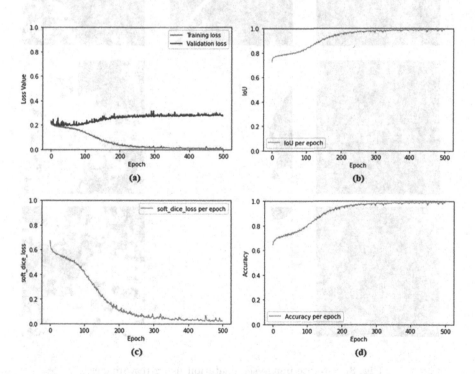

Fig. 2. Loss, accuracy, IoU and soft dice loss per epoch (a) Training and validation loss (b) Inter-section.

The modified U-Net model gives good values for IoU and soft dice loss. An Intersection over a Union value which is greater 50 is generally considered as a good prediction. It is evident that accuracy increase with an increase in

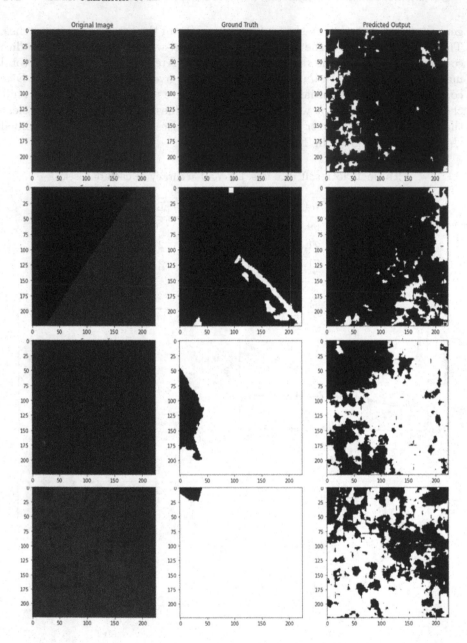

Fig. 3. Satellite image segmentation using test images

epochs. The model gives the best values for IoU on the training test which means this model is apt for flood-affected region detection in satellite imagery. The model detects the flooded regions from satellite images (see Fig. 3) with desirable accuracy. There is a significant decrease in soft dice loss per epochs.

In our case, the training loss is less than the validation loss (see Fig. 2). This shows that our model is fitting very precisely in the training data set but not in the validation data. It means that it's not generalizing accurately to unseen data. It is negligible since the primary concern is the value of IoU.

4 Conclusion and Future Work

A modified U-Net model is a good model for flood detection in satellite images. It will also good to implement different CNN architectures to analyse the model and compare the efficiency. It will also good to retrieve some meaningful information by processing with metadata and other data if available.

References

1. Nogueira, K., et al.: Data-driven flood detection using neural networks. In: MediaEval 2017, Dublin, Ireland (2017)
2. Said, N., et al.: Deep learning approaches for flood. In: MediaEval 2018, Sophia Antipolis, France (2018)
3. Lopez-Fuentes, L., van de Weijer, J., Bolaños, M., Skinnemoen, H.: Multi-modal deep learning approach for flood detection. In: MediaEval 2017, Dublin, Ireland (2017)
4. Bischke, B., Bhardwaj, P., Gautam, A., Helber, P., Borth, D., Dengel, A.: Detection of flooding events in social multimedia and satellite. In: MediaEval 2017, Dublin, Ireland (2017)
5. Zaffaroni, M., Rossi, C.: Water segmentation with deep learning models for flood detection. In: Proceedings of the 17th ISCRAM Conference, VA, USA (2020)
6. Ahmad, K., Konstantin, P., Riegler, M., Conci, N., Holversen, P.: CNN and GAN based satellite and social media data fusion for. In: MediaEval 2017, Dublin, Ireland (2017)
7. Bischke, B., Helber, P., Christian, S., Srinivasan, V., Dengel, A., Borth, D.: The multimedia satellite task at MediaEval 2017. In: MediaEval 2017, Dublin, Ireland (2017)
8. Ronneberger, O., Fischer, P., Brox, T.: U-Net: convolutional networks for biomedical image segmentation. In: Navab, N., Hornegger, J., Wells, W.M., Frangi, A.F. (eds.) MICCAI 2015. LNCS, vol. 9351, pp. 234–241. Springer, Cham (2015). https://doi.org/10.1007/978-3-319-24574-4_28
9. Mark, S., Andrew, H., Menglong, Z., Andrey, Z., Liang-Chieh, C.: MobileNetV 2: inverted residuals and linear bottlenecks. In: Proceedings of the IEEE Conference on Computer Vision and Pattern Recognition (2018)
10. Otsu, N.: A threshold selection method from gray-level histograms. IEEE Trans. Syst. Man Cybern. $9(1)$, 62–66 (1979)
11. Rezatofighi, H., Tsoi, N., Gwak, J., Sadeghian, A.: Generalized intersection over union: a metric and a loss for bounding box regression. In: Proceedings of the IEEE/CVF Conference on Computer Vision and Pattern Recognition (2019)
12. Dias, D., Dias, U.: Flood detection from social multimedia and satellite images using ensemble and transfer learning with CNN architectures. In: MediaEval 2018, Sophia Antipolis, France (2018)

13. Lopez-Fuentes, L., Farasin, A., Skinnemoen, H., Garza, P.: Deep Learning models for passability detection of flooded roads. In: MediaEval 2018, Sophia Antipolis, France (2018)
14. Kirchknopf, A., Slijepcevic, D., Zeppelzauer, M., Seidl, M.: Detection of road passability from social media and satellite images. In: MediaEval 2018, Sophia Antipolis, France (2018)
15. Said, N., et al.: Deep learning approaches for flood classification and flood aftermath detection. In: MediaEval 2018, Sophia Antipolis, France (2018)
16. Gerald, W.J., Bisshop, W., Senarath, U., Stewart, A.: Methodology for mapping change in woody landcover over Queensland from 1999 to 2001 using Landsat ETM+ (2002)
17. Ogashawara, I., Curtarelli, M.P., Ferreira, C.M.: The use of optical remote sensing for mapping flooded areas (2013)
18. Bischke, B., Borth, D., Schulze, C., Denge, A.: Contextual enrichment of remote-sensed events with social media streams. In: Proceedings of the 24th ACM International Conference on Multimedia (MM 2016). Association for Computing Machinery, New York (2016)
19. Sébastien, O.: Building segmentation on satellite images (2018)
20. Ticehurst, C.J., Dyce, P., Guerschman, J.P.: Using passive microwave and optical remote sensing to monitor flood inundation in support of hydrologic modelling. In: 18th World IMACS/MODSIM Congress, Congress, Cairns, Australia (2009)
21. Klemas, V.: Remote sensing of floods and flood-prone areas: an overview. J. Coastal Res. $J31(4)$, 1005–1013 (2015)
22. Schnebele, E., Cervone, G., Kumar, S., Waters, N.: Real time estimation of the Calgary floods using limited remote sensing data. Water $6(2)$, 381–398 (2014)
23. Liebrecht, C.: PDF hosted at the Radboud Repository of the Radboud University Nijmegen the perfect solution for detecting sarcasm in tweets# not. In: Proceedings of the 4th Workshop on Computer Approaches to Subject Sentiment Social Media Analysis (2014)
24. Ahmed, A., Sinnappan, S.: The role of social media during Queensland floods: an empirical investigation on the existence of multiple communities of practice (MCoPs). Pac. Asia J. Assoc. Inf. Syst. $5(2)$, 2 (2013)

Blockchain

Stockholm

Blockchain Based End-To-End Tracking System for COVID Patients

Surekha Thota$^{(\boxtimes)}$ ⓘ and Gopal Krishna Shyam ⓘ

School of Computer Science and Engineering, REVA University, Bengaluru, India
surekha.thota@reva.edu.in

Abstract. COVID-19 pandemic had disrupted the health, social and economic stability throughout the world. Centers for Disease Control and Prevention (CDC) had proposed various strategies to minimize the spread of virus; the government is implementing these strategies to take good care of victims. Depending on the severity of symptoms, COVID positive patients are either home quarantined or admitted to the hospital. Two essential facilities that ensure a hassle-free journey for COVID patients are i) publicly available transparent hospital management system that publishes the availability of COVID beds. ii) a transparent list of suppliers that delivers food and medicines to the home quarantined patients. Our proposed model i) builds an end-to-end tracking system for COVID patients ii) enables patients to choose their preferred hospital by checking the availability of beds and iii) guarantees a fraud free vaccination system.

Keywords: COVID · Coronavirus · Blockchain · Hospital tracker

1 Introduction

COVID-19 pandemic has taken the lives of many people all over the world. Millions of people are infected globally with coronavirus; the infection rate is increasing exponentially. This pandemic has not only disrupted the health but also left uncertainty in the world's economy. Researchers are identifying ways to stop this prolonged uncontrollable spread of the virus. They are trying to discover the vaccine to protect the human body against coronavirus. Centers for Disease Control and Prevention (CDC) and the government are enforcing various strategies to minimize the spread of the virus.

Some of these strategies are implementing strict lockdowns, maintaining social distancing, promoting regular handwash, and mandating people to wear masks, closing infected areas by indicating them as red zones. Implementing these precautionary measures could delay the exponential rise of infected cases. This delay gave sufficient time for the government to create the following facilities:

- manufacture COVID-19 test kits and PPE kits
- build rehabilitation centres
- increase hospital beds and ventilators

© IFIP International Federation for Information Processing 2021
Published by Springer Nature Switzerland AG 2021
V. Krishnamurthy et al. (Eds.): ICCIDS 2021, IFIP AICT 611, pp. 177–186, 2021.
https://doi.org/10.1007/978-3-030-92600-7_17

India is the second-largest populated country in the world. In such a highly populated country, it is difficult to provide treatment for all the COVID victims. The following are some of the challenges faced by the government, hospitals, and the public.

(a) Challenges faced by the Government

- Lack of Facilities: During the initial days of COVID, the country struggled with insufficient COVID-19 test kits, PPE, sanitizers, masks, hospital beds, and ventilators. The situation had improved drastically with time.
- Lockdown:

  – The economy was deprecated as many businesses were closed due to lockdown.
  – Unemployment due to shutdown of educational institutions, shopping malls, and cinema halls etc.
  – Migrant workers moved to their natives during the lockdown and hence industries struggled due to the unavailability of workers.

(b) Challenges faced by Hospitals

- Doctors, nurses, and hospital staff are at high risk
- COVID patient's inflow is more compared to patients who are getting discharged

(c) Challenges faced by people

- During the lockdown, the government and NGOs distributed subsidized food to the daily wagers and people below the poverty line. Still, they faced a lot of economic distress as they could not go to their work.
- Due to the sudden loss of jobs, people who want to go back to natives faced transportation issues.
- Employers are battling for existence, whereas employees are suffering from job loss or salary deduction
- Patients with mild symptoms are home quarantined and need to have a supply of necessities like milk, medicines, and food.

Patients with moderate to severe symptoms, elderly, people with other ailments like diabetics, blood pressure need to be admitted to hospital. Governments are trying their best to accommodate and provide treatment for all the people who have moderate to severe symptoms. Initially, there was no transparency in the vacancies of beds and ventilators; patients were going from one hospital to another in search of the hospital bed. Later the situation is improved with the COVID positive hospital bed management system (CHBMS) [6].

Our paper aims to propose a solution that considers the user's preferences when multiple vacancies are available for hospital beds. It also tracks end-to-end details of COVID patient.

2 Literature Review

It's a well-known fact that the COVID-19 infected cases will be increasing till the right medication and vaccination are discovered. India is the second highly populated country in the world, it is challenging to provide a medical facility to 1.38 Billion people. In India, as of today, there are two ways a COVID patient can get admitted to the hospital.

1. Private quota – Patients can directly contact the hospital and get admitted. Medical expenses for treatment have to be borne by the patient.
2. Government quota - Patients get admitted through Government referral. As per circular dated 4th of August 2020 from Commissionerate health and family welfare services, Government of Karnataka, all the referrals from BBMP (Bruhat Bengaluru Mahanagara Palike) admissions are done through COVID positive hospital bed management system (CHBMS) [6, 7].

The protocol for filling the Government quota seats are as follows:

- On validating and approving the details of the patient, the special commissioner (Health) shall block the bed in the CHBMS portal preferably in Government Medical Establishment.
- The information on allocation of beds will be intimated to patients and hospital for further action.
- The admissions and discharges through government referral should be recorded through CHBMS [6, 8]. The dashboard of bed availability is shown in Fig. 1.

Domain and technology go hand in hand. Blockchain's greatest characteristic stems from the fact that its transaction ledger is open to all. Hence, we can apply blockchain to provide an end-to-end tracking system for COVID patients. Research is being carried out on how blockchain technology can help to ease the administration during COVID pandemic. The framework proposed in [1] detects the unknown infected cases and predicts the contagion risk. It aims in building a blockchain-based P2P-Mobile application for helping the citizens to detect the sources of contagion in a crowd.

Blockchain-based infectious disease reporting system built-in [2] eliminates intermediaries and accelerate the process. It removes the possibility of data falsification and provides transparency to the monetary donation system. The author of [3] tries to explore the medical applications powered by blockchain and how these trustless systems can resolve the tension between maintaining the privacy and addressing public health needs in the fight against COVID-19.

Paper [4] proposes an architecture by integrating blockchain and AI-specific for COVID-19 fighting. SWOT analysis on blockchain-based prediction model in healthcare is carried out in [5]. It assures that blockchain plays its role to improve COVID-19 safe clinical practice. For COVID-19 test-takers [9] implements immunity certificates and digital medical passports (DMP).

The existing hospital tracking systems are confined to a specific hospital in a region or state. But the proposed solution is a global tracking system through which patients

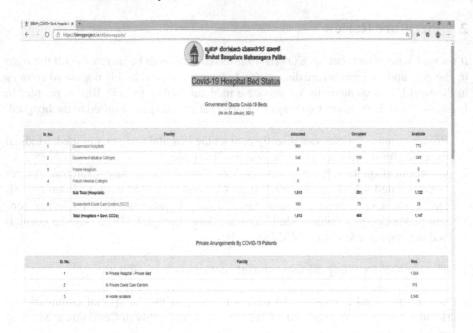

Fig. 1. COVID-19 hospital bed status – BBMP

can easily track the availability and look for nearest possible hospitals. It also helps to connect various service providers and patients.

In this paper, we take a step ahead to propose a blockchain based COVID Hospital tracker that provides transparency between the hospitals, medical stores, food suppliers, government, municipal corporations, and patients. Blockchain adds an unprecedented layer of accountability, holding patients, Government & Hospitals so that they act with integrity towards community. This immutable, decentralized and tamper proof nature of blockchain makes it safe from falsification of information and attacks. This considers patients preference, especially when multi-vacancies are available in various government hospitals.

3 Methodology

This paper proposes a blockchain-based transparent health care facility for people who seek hospital admission. We aim to build a blockchain based COVID Hospital tracker to ensure transparency between the hospitals, government, municipal corporations, and patients. In this proposed model, the government, hospitals, municipal corporations, and patients are various stakeholders and they represent different nodes in the blockchain. We track the patient's journey right from visiting test centers till vaccination. This system ensures transparency in displaying the bed availability by hospitals and eliminates any fraud or falsification of data. It not only verifies the authenticity of hospitals but also provides opportunity for the patients to choose his hospital of interest for treatment. The sequence diagram of the proposed methodology is represented in Fig. 2.

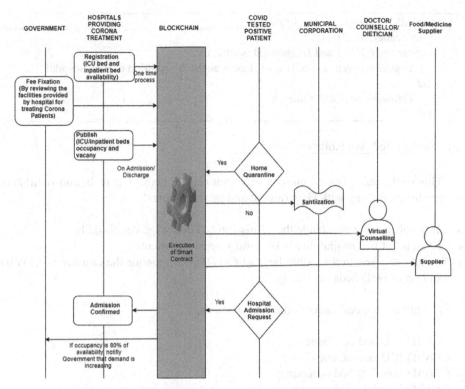

Fig. 2. Blockchain based end-to-end COVID tracking system

The following are various modules present in our COVID Hospital tracker.

(a) Registration

Hospitals providing COVID treatment must register to this blockchain. The authenticity of hospitals is carried out by MOH (ministry of Health). We intend to make this solution feasible beyond COVID. Hence the COVID and non-COVID beds are tracked separately. At the time of registration, hospitals should provide the total number of.

- COVID-ICU beds
- COVID-Inpatient beds
- Non COVID-ICU beds
- Non COVID-Inpatient beds
- Ventilators

The below algorithm 1 depicts the sequence of steps involved in registering hospitals to the blockchain.

Algorithm 1: Hospital Registration

Input : caller, *MOHEA*
if caller ==*MOHEA* and *HospitalID* is valid then
    Trigger an event to notify all listeners about the addition of new hospital
else
    Throw an error and rollback
end

(b) Publish Bed Availability

This model publishes the number of vacant general beds, ICU beds, and ventilators at each hospital along with Government and private Quotas.

- All hospitals have to provide the admission and discharge details daily
- The status of bed availability is kept transparent to patients
- To make this solution feasible beyond COVID, we monitor the number of COVID and non-COVID beds separately.

Hospitals must daily submit the below details.

- COVID-ICU bed occupancy
- COVID-ICU bed vacancy
- COVID-Inpatient bed occupancy
- COVID-Inpatient bed vacancy
- Non COVID-ICU bed occupancy
- Non COVID-ICU bed vacancy
- Non COVID-Inpatient bed occupancy
- Non COVID-Inpatient bed vacancy
- Vacant Ventilators

The below algorithm 2 depicts the steps involved in publishing the hospital availability.

Algorithm 2: Publish Availability Hospital Update

Input : caller, *HospitalEA*
if caller == *HospitalEA* then
    Update the *COVID_ICU_bed_occupancy,COVID_ICU_bed_vacancy,*
    COVID_Inpatient_bed_occupancy, COVID_Inpatient_bed_vacancy,
    Non_COVID_ICU_bed_occupancy, Non_COVID_ICU_bed_vacancy,
    Non_COVID_Inpatient_bed_occupancy,
    Non_COVID_Inpatient_bed_vacancy, Vacant_Ventilators
else
    Throw an error and rollback
end

(c) Test Result tracking

- Each test kit is uniquely identified by a test kit serial number,
- This module tracks the patient's details with the test kit serial number and test results.
- If the patients are tested positive, the patient details are with health counselors, municipal corporators, medical and food suppliers.

The below algorithm 3 depicts the sequence of steps involved in patient registration and mapping patient to test kit. Whereas the algorithm 4 depicts the sequence of steps involved in test result tracking.

Algorithm 3: Patient - Registration

```
Input : caller, TestcenterEA
    if patient has valid PatientID then
        Map PatientID to TestkitID
    End
else
    Throw an error and rollback
end
```

Algorithm 4: Test result tracking

```
Input : caller, TestcenterEA
if caller == TestcenterEA then
    if patient == PatientID
        PatientID_Testresult = Testresult
        Trigger an event to alert the test results to the patient
        if PatientID_Testresult = "positive"
            Trigger an event to alert Municipal corporations, Doctor,
            counsellor or Dietician, food or medicine supplier.
        end
    end
else
    Throw an error and rollback
end
```

(d) Home Quarantine

- Patients who tested positive are home quarantined if they are pre-symptomatic or asymptomatic.
- Home quarantine information is shared with virtual counselors, pharmacies, dieticians, and municipal corporations to provide moral support to the infected people.

- During this quarantine period, they need home delivery of milk, groceries, vegetables, and medicines. So, details of the nearest suppliers are provided to them.
- The status of the home quarantine patients is tracked along with the symptoms they have and the medication they follow.

(e) Admission Requisition and allotment

- Patients having moderate to severe symptoms or other ailments may get admitted to the hospital.
- If the patient prefers admission through government referral. On validating the details of the patient, the special commissioner (Health) shall block the bed in Government Medical Establishment.
- If the patient prefers admission directly to a private hospital, he can visit the dashboard prepared in 3.2 and decide the hospital he wishes to admit.
- Based on the availability, hospitals confirm the request.

The below algorithm 5 depicts the sequence of steps involved in patient admission to hospital.

Algorithm 5: Admission to hospital
Input : caller, *PatientEA*
if caller == *PatientEA* then
    Patient_preference= HospitalID
    If *HospitalID* has vacant beds
        Allot bed from *HospitalID* to *PatientID*
else
    Throw an error and rollback
   end

(f) Notification
    In a particular city, if the bed occupancy reaches 80% of availability, then a notification can be sent to the government health officials about the rapid filling of seats so that they can plan for arranging more beds.
(g) Certification of discharge
    The recovery status is noted along with his quarantine period.
(h) Vaccination

In addition to end-to-end tracking, our model includes a transparent vaccination system. Individual wish to take vaccine can reserve the slot by specifying the choice of the vaccine. This Blockchain vaccination module eliminates frauds involved in enhancing your rank for vaccine eligibility. The patient's identity is mapped with the vaccine serial number, date of vaccination, and vaccine. While traveling overseas, this mapping acts as a passport for COVID vaccination.

4 Results

In this section, we discuss the implementation details and results of the proposed transparent end-to-end tracking system for COVID patients and a fraud-free vaccination system. It is implemented on an Intel (R) Core (TM) i5-1035G1 CPU@1.00 GHz, 8 GB RAM and 64-bit operating system. Smart contracts were coded in remix IDE using solidity programming language. These smart contracts are deployed in a private Ethereum blockchain network created using Ganache. To support payment on the network we have used Metamask wallet. Figure 3 depicts a sample tracker of a patient who got admitted to private hospitals and also the sample record for the vaccine tracking system.

COVID+ Tracker

Details of Patient
Unique ID :6541 1255 XXXX XXXX
Name : Bob

Details of Test Kit
Test Kit Serial Number :9876 5432 XXXX XXXX
Date of test : 25 December 2020
Test Location : PQR Hospital

Home Quarantine : No
Beginning of Quarantine period : 26 December 2020
End of Quarantine period : 12 January 2021

Admission Details
Private/Government Referral : Private
Unique hospital ID :6754 5234 XXXX XXXX
Hospital Name :ABC Hospital
Date of admission :27 December 2020
Date of discharge :04 January 2021

COVID Vaccine Tracking System

Details of person taking Vaccine

Unique ID :6541 1233 XXXX XXXX
Name : Alice

Vaccination Details

Vaccine Name: XYZ
Vaccine Serial Number: 1234 5678 9876
Date of vaccine: 06 January 2021
Front line Worker: No

Fig. 3. COVID + tracker and COVID vaccine tracking system

5 Conclusion

Today's Hospital Management systems are limited to a particular hospital and are not transparent. The proposed COVID hospital tracker powered by blockchain technology brings transparency among government, patients, hospitals, municipal corporations, food, and medical suppliers. As the complete process is automated, the processing is faster without delay. This proposed solution can work beyond COVID. We also have introduced a transparent vaccination system that eliminates fraud during vaccination.

As part of the future scope, we draw some conclusions, by analyzing and implementing AI and ML algorithms on the recordings of COVID symptoms, medicines used, and the recovery details.

References

1. Torky, M., Hassanien, A.E.: Article COVID-19 blockchain framework: innovative approach. arXiv preprint arXiv:2004.06081 (2020)

2. Chang, M.C.F., Park, D.S.: Article how can blockchain help people in the event of pandemics such as the COVID-19? J. Med. Syst. **44**, 1–2 (2020)
3. Khurshid, A.F.: Article applying blockchain technology to address the crisis of trust during the COVID-19 pandemic. JMIR Med. Inform. **8**(9), e20477 (2020)
4. Nguyen, D.F., Ding, S.M., Pathirana, T.P.N., Seneviratne, F.: Blockchain and AI-based solutions to combat coronavirus (COVID-19)-like epidemics: a survey. TechRxiv. Preprint. https://doi.org/10.36227/techrxiv.12121962.v1 (2020)
5. Fusco, A.F., Dicuonzo, G.S., Dell'Atti, V.T., Tatullo, M.F.: Blockchain in healthcare: insights on COVID-19. Int. J. Environ. Res. Public Health **17**(19), 7167 (2020)
6. Commissionerate health and family welfare services. https://covid19.karnataka.gov.in/storage/pdf-files/56152265447195541878.pdf. Accessed 4 Dec 2020
7. Covid Emergency Workflow. https://covid19.karnataka.gov.in/storage/pdffiles/Public%20Information/GI-Covid-19%20Emergency%20Work%20Flow.pdf. Accessed 12 Dec 2020
8. BBMP — COVID+ Govt. Hospital Bed Status. https://bbmpproject.in/chbmsreports/. Accessed 8 Jan 2021
9. Hasan, H.R., et al.: Blockchain-based solution for COVID-19 digital medical passports and immunity certificates. IEEE Access **8**, 222093–222108 (2020). https://doi.org/10.1109/ACCESS.2020.3043350

Decentralized Application Using Ethereum Blockchain on Performance Analysis Considering E-Voting System

B. Sriman$^{(\boxtimes)}$ ⓘ and S. Ganesh Kumar

Department of Computer Science and Engineering, SRMIST, Chennai,
Tamil Nadu, India
{sb7072,ganeshk1}@srmist.edu.in

Abstract. This paper investigated how the technology behind many of the cryptocurrencies and the Bitcoin is Blockchain Technology. Bitcoin Blockchain was the first blockchain technology in 2008 to provide as the public digital payment with no other third party or the middleman interface. The result showed that technology replaced the cash exchange method (i.e., Cryptocurrency, Digital Assets, and Distributed Ledger). Furthermore, the key main characteristics of blockchain and their potentials to technology (e.g., Decentralization, Immutability, Transparency, Persistency, Auditability, Security, and Privacy). Most of the application was built by using Blockchain Technology. Blockchain-as-a-Service (BAAS) for the enterprise (e.g., Ethereum, Cloud Storage, Supply Chain Communication, smart contract and E-Voting System). In this paper, we developed a democracy E-Voting System, that replaces the ancient mode of voting system. We tested and implemented our E-Voting system that runs with a smart contract by using the E-Wallets. After the voting is confirmed, it holds all the data of voters, ballots and transaction data. Performance analysis of the E-Voting System is theoretical and practical implications of the result are discussed.

Keywords: Bitcoin · Blockchain · Ethereum · Proof of work · Smart contract · E-wallets · E-voting

1 Introduction

Blockchain is a distributed decentralized application, where the nodes are connected through a peer-to-peer network and agreed to follow the rules that are mentioned in the smart contracts [1]. Where, the smart contracts [2] are the codes that are written in the solidity programming language. In our paper, we used the Ethereum blockchain platform to build a decentralized application for the e-voting system. In blockchain, we used a decentralized network instead of using the centralized network. In the peer-to-peer [P2P] network, all nodes share their records to every node in the distributed network [3]. A copy of one node's data will be shared with all the other nodes in the blockchain.

© IFIP International Federation for Information Processing 2021
Published by Springer Nature Switzerland AG 2021
V. Krishnamurthy et al. (Eds.): ICCIDS 2021, IFIP AICT 611, pp. 187–199, 2021.
https://doi.org/10.1007/978-3-030-92600-7_18

Here, no central authority is given to any of the nodes. As it is decentralized, anyone in the network can verify the other node whether it is valid or not. The node which is verified by the participant is called the miner. The miner is the participant, who undergoes for mining work. Here, mining is the process of solving mathematical crypto-puzzle [4]. Let us look deeply into the networks in which we are talking about. In the peer to peer [P2P] network [4], the shared data across the nodes are stored as the bunch of documents known as the blocks. Within, all nodes are together to construct the new registry publicly. Here, the data are secured under the cryptographic hash and are validated by using the consensus algorithm. In this paper, we are implementing our E-voting application to ensure that our vote is counted and secured [5]. As discussed above, the node ensures the sample of data beyond the web are the same. This is to the motive why we are constructing our E-voting System on the Ethereum blockchain [6]. In the existing system, when we use the web application, we will utilize a gateway to connect with the database among network and the data are shared through the central servers or the central database. If any transaction is encountered, the information will be shared through the central servers. So, our information may be accessed or been modified by any of the users. In voting system with the central servers, we may get a few problems as follows:

1. Mutability may occur in the database
2. The source code on the can be modified by any of the intruders

In our paper, we are going to build the E-Voting System which is decentralized and the nodes are connected to cast their votes. This is to ensure that our vote is casted and they are counted or not. Let us see the process deeply in the upcoming sections. Blockchain allows developers to build many of the Dapps by using different platforms. Here, we are going to build an application using the Ethereum plat- form [7]. One of the most important platforms to execute the arbitrary codes to run on the programs in Ethereum. With the help of smart contracts, the distributed infrastructure could be able to complete the different types of projects in Ethereum. Ethereum allows us to build the decentralized and the fault-tolerant applications [8], which removes the intermediaries and provides transparency to all the nodes in the network. A new tradable token can be created by using the Ethereum that can be used as the cryptocurrency.

The wallets that are compatible with the Ethereum blockchain are used as token coin as a standard API [9]. To build a blockchain-based organization, the nodes to be added in a network should be agreed for the constraints [10]. Smart contracts will execute the contracts automatically only when all the constraints are satisfied. A clear concept of Ethereum should be known before entering into it. Now, we are going to learn few things about the Blockchain – Ethereum platform and the working of smart contracts. Smart contracts are the codes to execute the Ethereum blockchain with the help of Ethereum Virtual Machine [EVM]. This is where all our business applications work. The smart contracts play an important role in reading, writing, and executing the data on the blockchain. The code is written in a solidity programming language, similar to the JavaScript [11]. The same type of things can be done by using these codes for different applications. The smart contracts on the blockchain are related to the resource on the web. Therefore, our smart contract communicates with the business logic that executes with all our data and then the database layer of the blockchain is represented by public ledger [12].

2 Preliminaries Work of E-Voting System in Etherum Blockchain

2.1 Motivations

- When we are traveling aboard or not available in our home town, the E-Voting system should be a traceable back end and transparent to the voter by using the Dapp in Ethereum platform.
- The E-Voting System uses a mobile app or computer to install the Dapp and the internet is plays major role in the whole voting procedure.

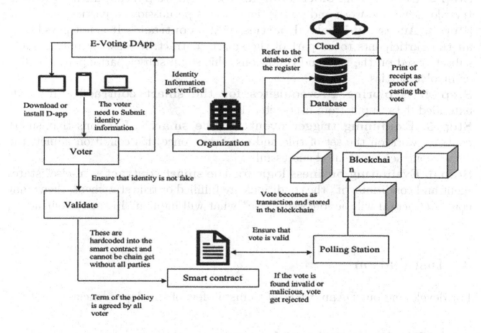

Fig. 1. Blockchain architecture for democracy E-Voting system

2.2 Challenges Faced by E-Voting System

- Possibility of manage activities by a hacker who might straight infect and hack the servers.
- The occurrences of constraints or to modify voting procedures can be large number [13].
- No method to prove if your vote has been added in the E-Voting System.

3   Research Methodology

3.1   Ethereum Smart Contract

A smart Contract is a digital-based contract or digital version of a contract. it acts as a per contract, which runs on the blockchain node and executes the requirement of parties. The smart contract asset of a parameter in blockchain to become a trustless network. If we design the right, a smart contract will surely be successful as it is driven by a computer system without human intervention.

3.2   Step to Design a Smart Contract in Ethereum Blockchain

Step 1. Understanding the Business problem – The Smart contract has set of parameter, it includes the business problem statement or understanding the business problem statement.

Step 2. Defining the blockchain network and its participants – Creating the blockchain network and giving them access permission to parties.

Step 3. Accessing involved parties of the contract – It is not possible to all the participants to be part of the smart contract. it must be only a small subset based on the rule and conditions, this small subset participant is called as involve parties.

Step 4. Exploring Prerequisites for the smart contract – It needs extended data, input from users, etc. [14].

Step 5. Examining trigger event for the smart – It means that smart contract work on the set of rule and condition, once this condition is met, the smart contract is executed successfully.

Step 6. Evaluating business logic for the smart contract – "if-else" statement and condition. "if" the conditions are fulfilled or true statement, then what course of action will be followed, "else" what will happen? In smart contracts.

4   Installation

For developing our DApp, we need to install few of the dependencies.

4.1   NPM (Node Package Manager)

The initial dependency is the npm that comes along with the node.js. Through the integration of npm, the truffle comes with standard. It means that we can distribute and use the contracts, Dapps and even more Ethereum enabled libraries through npm. To check the version of the node, we use the command: node -v.

4.2 Truffle Framework

The truffle framework [15] is our second dependency, which allows us to build the DApps in the Ethereum blockchain. It helps us to interact through the smart contract. We use the npm package to install the truffle. To install the truffle, we use the command: npm install -g truffle. Make sure the truffle is compiled and deployed before interacting with the smart contracts and then to interact with them by using the Web3. Here, we suggest that truffle The contract library is much easier and robust.

4.3 Metamask

The simplest way to interact with the DApps is by using the metamask. To connect with the Ethereum network, we use an extension in Chrome or Firefox without running the whole node on the browsers. The metamask is capable of connecting with the main network and any test networks like ropsten, rinkeby, kovan, or a local Ethereum development called Ganache or Truffle Develop.

4.4 Ganache

The local in-memory blockchain is our third dependency and it is called Ganache. It can be downloaded by using the website from the truffle framework. After installation, ganache gives 10 free accounts within our local Ethereum blockchain. This means, initially the original account holds 100 ethers. A solidity programming language is a language that is highly recommended for Ethereum blockchain. Here, we are using the sublime text and we have installed the Ethereum package which supports solidity language.

5 Implementation

5.1 Building the Ethereum Smart Contract

For a smart contract, the language used is the Solidity. The programming language that is developed on top of the Ethereum Virtual Machine [EVM]. The main purpose of solidity is to store all the states of the token, which are sent and received by the nodes. The programmers who are experienced in JavaScript, C++, Python can easily understand the solidity language. If the solidity language is known well, then the program for Ethereum smart contracts can be written easily. The election file is written with the solidity code. Initially, we declare the solidity language with its pragma version as required. Then, the name for our contract will be given by the contract name in the smart contract with the keyword "contract" [17]. The state variable of the candidate name will be stored and allows us to store it in the blockchain. We declared it as a variable and that variable is a string.

5.2 Setting up a Smart Contract in Solidity Language

Pragma version 0.6.0;

```
{
```
Step 1. Struct keyword keep track the number of candidate in an election
// Struct candidate
Step 2. Set the unique function identifier id for each candidate of various size
//Uint id;
//String name
Step 3: Number of vote count to the candidate
// Uint votecount;

```
}
```

This string will be visible publicly in the blockchain. Because it allows us to access the value of our smart contract. The function is a constructor and it is known by the solidity codes. Now, we have created our base for our smart contract. This is how we will code for the candidate model and it is given below as solidity code for storing multiple candidates and attributes.

5.3 Code Snippet Mapping the Candidate

Step 1: Calling the function Election contract contain a persistent data in state variable and function
// Contract Election {
Step 2: set the function identifier id
// struct candidate
Uint id;
String name ;
Uint votecount;
}
Step 3: Mapping is seen as hash tables which refers the type as arrays and struct to read and write data as mapping (uint => candidate) Public view candidate. Here, we modified the code for a candidate in the solidity. The solidity allows us to create our structure type as we have already done in our code. By declaring the struct, we can't simply give the candidate [16]. First of all, we should instantiate it and assign the variables, before we write the codes to store it in the blockchain. The next step is to store the candidate. For that, we need some space for structure type. We can simply do solidity by mapping for this. Here, mapping is to resemble an array or the hash, that associates with the key pair values. And the Solidity code for mapping a candidate structure, unsigned integers are explained.

5.4   Code Snippet Storage the Candidate Count

Step 1: Calling the function state variable and function
// Contract Election {
Step 2: set the function identifier id
// struct candidate
Uint id;
String name;
Uint votecount;
}
Step 3: Mapping is seen as hash tables
// mapping(uint => candidate) Public view candidate
Step 4: Possible to store UTF-8encoded data in string variable Candidate count
Uint Public candidatecount:

}

5.5   Code Snippet Using Counter Cache

Step 1: Calling the function state variable and function
// Contract Election {
Step 2: function add Candidate or remove the candidate
Remove lsit candidate name from array and struct
function addcandidate (srting name) private
{
Candidat count++
Step 3: number responses every time are display a counter cache keep a separate
counter
Candidate(camdidatescount, _name,0):
}
}

To write data in the blockchain, we assign the key-value pairs to the state variable for mapping. To get a better function, we set the function as previously with the candidate name in the test. Let's how the candidate's vote are gets counted in solidity code for reading/write, store candidates in the codes.

5.6   Testing Smart Contract in Ethereum Main Network

Now, we are going to test our E-voting smart contract in the ethereum main network by using the Metamask and Ganache. We need to create the new test file in the command line from the root of our paper by using the following command: touch test/election.js To bundle the truffle files, we are using the JavaScript file with the mocha testing framework. Here, we write the JavaScript file to simulate the client-side interaction with a smart contract. The truffle test command is given to test that our contracts are running perfectly or not. We use the Client-side application to interact with the smart contracts. Let us see how it works and how it is helpful for us. In the truffle package, we get the HTML and the JavaScript file. We just want to modify for our paper. Here, these files keep us to write any of the CSS as we wish to modify. We use the lite server in our paper to develop our e-voting decentralized application. For all these, we no need to worry about the codes. It will keep the HTML and JavaScript codes simple and focuses on it. Let us now focus on how to develop smart contracts as a part of our decentralized application.

1. Web3 is a JavaScript library that allows us to interact with the client-side application to the blockchain. We configured the Web3 inside a function as "initWeb3"
2. Initialize the contracts: we have to initialize some value inside the contract function to interact with it.
3. Render function: from the smart contract, it renders all the function layouts on the content page with the data.

We have created a smart contract to list the candidates. Now, looping is passed through each candidate in the mapping and rendering it to the table and the current account is also fetched and displayed on the page. Initially, we have to check whether we have migrated our contracts through truffle or not. To do migrate function, we should use the command truffle migrate.

5.7   Lit Server – Installation and Configuration

we start our web server and start using DApp [17]. The lite server.

Initiating the webserver
Open the bs-config.json in the text editor and examine the contracts
```
{
"server": {
"baseDir":["./src", "./build/contracts" ]
}
}
```

The above codes tell the Lite-server to which and where the files to be included in the directory. Here, we added ./src directory for the website files and contracts, we added ./build/contracts directory. The dev command to check script in package.json file in the root directory of our paper. The script code looks as follows:

5.8   Code Snippet for Testing the Server

```
"script": {
"dev": "lite-server",
"test": "echo " error: no test specified " && ext1"
}
```

The above script code tells the npm to run the lite server. When we give the command "npm run dev", the server starts running and the local host for our paper will be displayed. After the npm command [18], it should automatically open the web browser with our client-side application. Here, the core functionality is to increase the candidate's vote count by reading the struct out of the "candidates" mapping and increasing the "vote count" by 1 with the increment operator.

Two things needed to be tested:

1. Whether the function increments the candidate's vote count or not
2. Whether the voter is added to the mapping whenever they vote or not

Now, we can ensure that if any transaction during the vote fails, when an error message is returned. In that message, we are able to revert that the error message contains the "revert" substring. Later, we could finalize that our contract state is unaltered and that the candidate who participated didn't receive any vote. To prevent double voting, we ensure to write the test code. Now, we query for the candidate id in the form. It will go to the smart contract and pass the id to call the vote. This is called an "asynchronous call". When it gets over, the page will be shown as the content to the user. After submitting, we will be directed successfully to casted a vote from the page. Then, ensure that the lite server is running. Because our assets are being served to the browser and proceeded to the browser page. There, we see the list of candidates to vote, which are connected to the ethereum blockchain main network. And it will be connected by the client-side application.

A popup message will come from the Metamask to confirm that we want to submit the transaction to the ethereum blockchain and to cast our vote. From the metamask account address, some amount of gas fee, a gas limit will be charged. It will be equal to 2.16 USD dollars. After submitting the transaction, the user will be able to cast their vote and submit it successfully on the Ethereum blockchain main network. Finally, we have successfully casted our vote with the client-side e-voting in the distributed application.

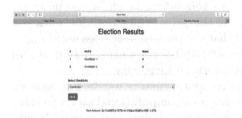

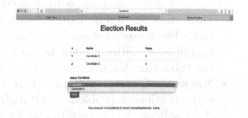

Fig. 2. Screenshot for election results

Fig. 3. Screenshot for client-side interact in E-Voting system

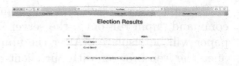

Fig. 4. Screenshot of metamask confirmation

Fig. 5. Dapp vote successfully

6   Discussion

- **Performance analysis:** As per the literature survey of the blockchain voting systems listed below are completely 13 s faster than the traditional voting systems. For multiple number of transactions, the same blocks are being created by inconsistent system process. Due to high scalability issues, only few of the nodes were tested across the distributed network with 10 nodes as shown in the Fig. 6.

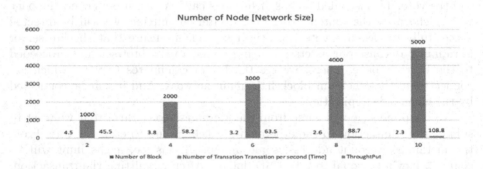

Fig. 6. Number of TPS vs network size

- **Measured Throughputs:** The throughput is analyzed by comparing it with the network size, number of transactions per second (in time). Based on the network size, the time will be needed to generate a block in the network. More than the public Ethereum network, the private Ethereum network has faster block propagation. It is observed that the performance does not changes significantly with different network sizes for POW consensus. It is shown in the table that the observed output drift with an increasing.
- **Network Configuration:** Where the computational power in the network is the bottleneck problem and it highly requires computational power to solve the mathematical crypto puzzle for verifying the transactions around 10 min and add the blocks into the blockchain.
- **Size of the network:** Due to the limitations in the network like size of a block, number of blocks and the transactions per second are some of the restrictions for the performance that are to be reduced.

Table 1. Measured throughput

Number of node [Network size]	Number of block	Number of transaction per second [Time]	Throughputs [Tps]	Measured throughput
2	4.5	1000	45.5	4.5 (89%)
4	3.8	2000	58.2	5.8 (98%)
6	3.2	3000	63.5	6.3 (75%)
8	2.6	4000	88.7	8.8 (70%)
10	2.3	5000	108.8	10.8 (66%)

- **Scalability/Throughput:** Speed of the network. The time is taken to do the transaction and mined it onto the network. In the Ethereum, it takes 10 min to confirm the transaction in the network [19]. Because of our transaction in the network wait up till 6 more blocks to be added in the network. So, throughput is a big problem in the Ethereum. The number of Transactions Per Second (TPS) in Ethereum is much lower than 2000 Transaction Per sec (TPS) for other payment alternatives like Visa take 21000 Transaction Per Sec (TPS).
- **Cost:** As we saw earlier, the mining process in Ethereum is expensive and the variable value of the Gas price adds to the trouble. The Gas price is unpredictable and suddenly the price of Gas will increase because of the volatile consumer building level of production application on Ethereum extreme challenge.
- **Decentralization:** In Bitcoin, 5 miner groups get together to control over 51% of the network. This violates the base of the Blockchain, which is decentralized because of these 5 miner groups can together administer or control the entire system. The same can happen in Ethereum. Though the algorithms for mining were designed in such a way that the chances of this happening are less.
- **Size:** The size of the Ethereum Blockchain is increasing, and nodes have to be highly equipped to store Blockchain of this size.
- **Ethereum:** Based DApps are not user-friendly yet.

7 Conclusion

In this paper, we developed a dapp as novel and optimistic web application so called "blockchain application in E-Voting System". We conclude that the main goal of our dapp development is to do E-Voting System Securely [20] by using the Ethereum blockchain in the main network. We successfully implemented our dapp for voting to the candidates transparently by the voters and in a transparent and verifiable manner. For future work, the time complexity and the constraints for the node scalability of the users to cast their vote from the registered network can be improved efficiently also the scalability issues.

References

1. Nakamoto, S.: Bitcoin: a peer to peer electronic cash system. https://bitcoin.org
2. Buterin, V.: Ethereum white paper: a next-generation smart contract decentralized application platform (2013). https://github.com/Ethereum/wiki/wiki/White-Paper
3. Wood, G.: Ethereum: a secure decentralized generalized transaction ledger. Ethereum Proj. Yellow Pap. **151**, 1–32 (2014)
4. Kshetri, N., Voas, J.: Blockchain-Enabled E-Voting. IEEE Softw. **35**, 95–99 (2018). https://doi.org/10.1109/MS.2018.2801546
5. Li, M., et al.: CrowdBC: a blockchain-based decentralized framework for crowdsourcing. IEEE Trans. Parallel Distrib. Syst. **1**(6), 1251–1266 (2018). https://doi.org/10.1109/TPDS.2018.2881735
6. Dhulavvagol, P., Desai, A., Ganiger, R.: Vehicle tracking and speed estimation of moving vehicles for traffic surveillance applications, pp. 373–377 (2017). https://doi.org/10.1109/CTCEEC.2017.8455043
7. Shafeeq, S., Alam, M., Khan, A.: Privacy-aware decentralized access control system. Future Gener. Comput. Syst. **101**, 420–433 (2019). https://doi.org/10.1016/j.future.2019.06.025, http://www.sciencedirect.com/science/article/PII/S0167739X18332308
8. Counterparty (2019). https://counterparty.io/. Accessed 30 Nov 2019
9. Cruz, J.P., Kaji, Y.: E-voting System Based on the Bitcoin Protocol and Blind Signatures (2016)
10. Cruz, J.P., Kaji, Y., Yanai, N.: RBAC-SC: role-based access control using smart contract. IEEE Access **6**, 12240–12251 (2018)
11. Democracy.earth (2019). http://democracy.earth/. Accessed 30 Nov 2019
12. Di Francesco Maesa, D., Lunardelli, A., Mori, P., Ricci, L.: Exploiting blockchain technology for attribute management in access control systems. In: Djemame, K., Altmann, J., Bañares, J.Á., Agmon Ben-Yehuda, O., Naldi, M. (eds.) GECON 2019. LNCS, vol. 11819, pp. 3–14. Springer, Cham (2019). https://doi.org/10.1007/978-3-030-36027-6_1
13. Di Francesco Maesa, D., Mori, P., Ricci, L.: Blockchain based access control. In: Chen, L.Y., Reiser, H.P. (eds.) DAIS 2017. LNCS, vol. 10320, pp. 206–220. Springer, Cham (2017). https://doi.org/10.1007/978-3-319-59665-5_15
14. Di Francesco Maesa, D., Mori, P., Ricci, L.: Blockchain-based access control services. In: 2018 IEEE International Conference on Internet of Things, iThings, and IEEE Green Computing and Communications, GreenCom, and IEEE Cyber, Physical and Social Computing, CPSCom, and IEEE Smart Data, SmartData, pp. 1379–1386. IEEE (2018)
15. Di Francesco Maesa, D., Mori, P., Ricci, L.: A blockchain-based approach for the definition of auditable access control systems. Comput. Secur. **84**, 93–119 (2019)
16. The European Parliament and the Council of the European Union, regulation (EU) 2016/679 of the European Parliament and of the council of 27 April 2016, Off. J. Eur. Union, 4.5.2016
17. The path to self-sovereign identity (2019). http://www.lifewithalacrity.com/2016/04/the-path-to-self-sovereign-identity.html. Accessed 30 Nov 2019
18. Secure.vote (2019). https://secure.vote/. Accessed 30 Nov 2019

19. Yu, S., Lu, K., Shao, Z., Gou, Y., Zhang, B.: A high-performance blockchain platform for intelligent devices. In: IEEE International Conference on Hot Information-Centric Networking (2018)
20. Domingue, J., Bachler, M., Quick, K.: Smart blockchain badges for data science education. In: 2018 IEEE Frontiers in Education Conference (FIE) (2018)

Enhanced Privacy Protection in Blockchain Using SGX and Sidechains

M. Mohideen AbdulKader$^{(\boxtimes)}$ and S. Ganesh Kumar

SRM Institute of Science and Technology, Chennai, India
{mm4123,ganeshk1}@srmist.edu.in

Abstract. Blockchain is a peer to peer network that is decentralized in nature. It has immutable and persistent property which make this technology more secured. Blockchain is not only suited for crypto currencies but also used for many applications like identity protection, smart contracts, health care, online polling system and much more. In addition, blockchain network holds a distributed ledger which makes all data available to every node in the network. Due to this, security and privacy protection becomes a major concern in blockchain technology. Some of existing solutions to blockchain privacy issues are homomorphice encryption, zero knowledge proofs, ring signature and multiparty computations. Though existing privacy preservation mechanisms secures the blockchain network from privacy leakage. It has limitations when implemented in large scale applications. In this paper, a three layered protection scheme is proposed to enhance the privacy of blockchain technology. This idea of three layered protection integrates randomized address generation, content erasure mechanism and IntelSGX together and forms a secure architecture. It will enhance the security and privacy of blockchain network to a greater extent.

Keywords: Blockchain privacy · Randomized address generation · Cryptography · Content erasure mechanism · Blockchain security · Anonymity · Distributed ledger technology

1 Introduction

Blockchain is a decentralized network and append only digital distributed ledger technology. Blockchain network is secured by cryptographic algorithms which enables transfer of digital assets between the nodes. With the use of blockchain application Bitcoin was the first digital cryptocurrency built over it which was introduced by Satosi Nakamoto in 2008. Bitcoin is an open source application that allows any number of users to join the network. By joining the blockchain network these nodes can make transactions of digital currency among them [31]. Based on the characteristics and usage blockchain can be categorized into public and private blockchain. Even though blockchain is considered as one of the secure technology there exists some privacy concerns that needs to be addressed (Fig. 1).

© IFIP International Federation for Information Processing 2021
Published by Springer Nature Switzerland AG 2021
V. Krishnamurthy et al. (Eds.): ICCIDS 2021, IFIP AICT 611, pp. 200–209, 2021.
https://doi.org/10.1007/978-3-030-92600-7_19

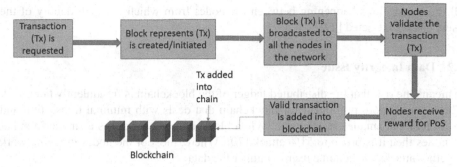

Fig. 1. Blockchain transaction architecture

The above figure represents the transaction that happens in a blockchain network. Initially a transaction Tx is requested in the blockchain network. Block representing the requested transaction Tx are created. This created block is broadcasted to all the participants in the network. All the nodes validate the transaction in the network by undergoing consensus mechanisms. If the transaction Tx becomes valid in it is added as a block in the blockchain. Blockchain can be classified majorly into public and private based on their characteristics and usage. Public blockchain is an open source network where any users can join that as a participant node. It have permission less access to the network and uses proof of work or proof of stake consensus mechanisms to validate the transactions. Whereas, private blockchain restrict access to only authorized users. Identity of the user is known in the private blockchain. Most private blockchain network uses proof of authority as consensus mechanisms to validate the transactions. Both public and private blockchain network holds a distributed ledger where all the information related to the transactions are stored which are visible to all other participants of the network. This makes the technology prune to security and privacy attacks [30]. In this paper, privacy challenges faced by the blockchain are analyzed based on which a three layer protection mechanism is proposed. It integrates three major methods such as randomized addressing scheme, content erasure mechanism and IntelSGX together to improve the security and privacy issues on blockchain network.

2   Security and Privacy Issues on Blockchain

Blockchain faces some major security and privacy related issues which are set backs the growth of this technology which are described as follows.

2.1   Transaction and Identity Privacy Issue

Transaction related information and data are stored in the distributed ledger of the blockchain which is visible to all other nodes in the network. This scenario leads to privacy issues. An attacker node can make use of transaction graph analysis to identify and extract the transaction information from the ledger [1]. Sensitive information such as user identity and information on transaction amount can be extracted out by this technique. Even though anonymous address was assigned to the users, an attacker can trace all the transactions happening between the nodes from which original identity of the users can be extracted [29].

2.2 Data Integrity Issue

It means the risk that the distributed ledger of the blockchain is fraudulently tampered. This issue is more prominent for blockchain that deals with minimal transaction and uses chains of smaller storage size. When blockchain network uses a smaller number of nodes then it is prone to 51% attack [20]. Where, most of the nodes in the network together attacks the genuine user to tamper the data.

2.3 Data Confidentiality Issue

It means the risk that sensitive information are exposed to all participants in the blockchain network. Anonymous address can be assigned to the users in the blockchain to avoid the leakage of user identity information [21]. Even then, in the case of private and permissioned blockchain achieving user anonymity is quite difficult.

2.4 Issue on Availability

It means the risk that participants cannot access blockchain network due to Denial of Service (DoS) attack. Availability issue depends on the number of nodes and transactions happening in that blockchain network. Private chains are generally smaller in size and easily disrupted by traditional Denial of Service (DoS) attack [22]. In addition, blockchain also faces interoperability risks when two different blockchain network need to operate together.

3 Privacy Preservation Mechanisms and Its Limitations

3.1 Bitcoin Privacy by Mixing Services

Mixing mechanisms are mainly implemented on bitcoin applications to improve the privacy. This mechanism was proposed by Chaum and it is used in bitcoin transactions [2]. Bitcoin transaction means that two users in a blockchain network transfers a digital currency between them without the use of any central server node [6]. Unlike normal transaction bitcoin does not use any central server node. Between two transacting node an intermediate node is added in this technique. Initially, the sender node sends the transaction amount in form of bitcoins to the intermediate node. Intermediate node collects the bitcoins sent from different sender nodes [5]. Collected bitcoins are then transferred to the respective receiver node. By this method, the sender and receiver of the transaction are kept anonymous [3]. An attacker node cannot identify the exact details of the transactions happened in the network (Fig. 2). This bitcoin mixing services are of two types such as mixing with central node and mixing without a central node [4].

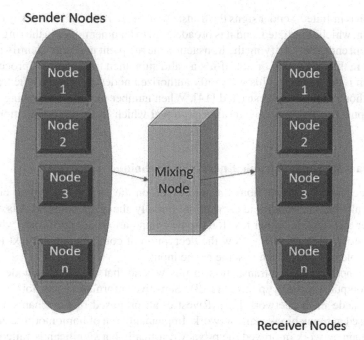

Fig. 2. Bitcoin mixing services (Source: D.Ron et al. 2013)

3.2 Privacy by Zero Knowledge Proof Mechanism

Zero knowledge proof mechanism is one of the widely used technique in blockchain network to improve the privacy which is proposed by Goldwasser et al. [10]. In this technique, in order to make a transaction the sender node need to prove a statement to be true to the recipient node. If the statement is proved then it can initiate the transaction. In addition, while proving the statement no additional information should be revealed to the recipient. Transactions will be performed only after successful verification of the proof by the recipient. There are two types of zero knowledge proofs that is interactive and non-interactive zero knowledge proof mechanisms. In both the mechanisms it is difficult for the attacker to extract out the sensitive information. Interactive zero knowledge proof allows two-way communication between the sender and receiver. Whereas, in non-interactive method there is no two-way communication between the sender and recipient nodes.

3.3 Privacy by Signature Scheme

Digital signature scheme named as ring signature is proposed by Rivest et al. In this mechanism, n number of nodes form a group called as a ring. In this ring, n number of transactions happen within the ring nodes. After a successful transaction within the ring group that transaction is added as a block in the blockchain network. In ring signature scheme there is no trusted third-party node. Transactions are processed in a decentralized way without the use of central or third-party node [3]. In this technique, when the transaction is initiated, sender signs the transaction. Sender uses its private key to sign the transaction, which is initiated, and it is broadcasted to all other nodes in the ring group. In the recipient end, after verifying the transaction the recipient node can identify the signer is present in the ring group or not. If it's a valid sign, then transaction is processed else transaction is declined. By this way, only authorized nodes can process the transaction and unauthorized users are restricted [14]. When number of nodes in the ring increases then computation also increase in a proportional which leads to computational errors [26].

3.4 Privacy by Homomorphic Encryption Technique

In this mechanism, computations are performed on the already encrypted cipher text which is called as homomorphic encryption. Initially, the given plaintext A is encrypted into cipher text f(A). In cipher text f(A) some computations are performed which gives the computed cipher text f(B). Now the decryption of computed cipher text f(B) gives the output plain text A which is same as the input.

Sender node sends the transaction in this way so that the receiver node gets only the final output after decryption (Fig. 3). Sensitive information are not disclosed to any other node in the network [23]. Rivest et al. proposed this mechanism which is majorly used in many blockchain network. Implementation of homomorphic encryption in blockchain network improved the privacy drastically but significantly failed for large scale blockchain applications [28].

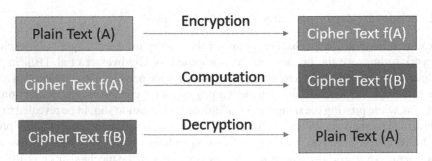

Fig. 3. Privacy by homomorphic encryption

3.5 Privacy by Multi-party Computation

In this mechanism, computations are performed by multiple nodes together without revealing their original identities. Multi-party computations are majorly used by blockchain smart contracts as it involves multiple nodes to sign a contract. It does not involve any third party node and is completely performs the computations in a decentralized way [25]. The major highlight of this technique is that sensitive data are divided into multiple segments. Each segments of sensitive data are distributed and stored in different nodes of the blockchain network. It enhances the privacy and security of the sensitive information to a greater extent [31]

4   Limitations of Existing Blockchain Privacy Mechanisms

Existing mechanisms such us bitcoin mixing services, signature scheme, multi- party computation, zero knowledge proofs and homomorphic encryption are widely used in many potential blockchain application [16]. These mechanisms delivered a considerable protection to sensitive transactional data. Even then it has several limitations which was discussed in the below table. Based on this, new methodology for privacy preservation on blockchain is proposed (Table 1).

Table 1. Limitations of existing privacy mechanisms

S. No	Privacy mechanism	Applications	Limitations
1	Mixing services	Mixcoin Blindcoin CoinShuffle Tumblebit	Prone to Denial of Service (DoS) attack Third Party Node (TTP) may disclose sensitive information
2	Zero knowledge proof	Zerocoin Zerocash	High cost for computing the proof Usage of storage space is very high
3	Homomorphic encryption	Confidential transaction Paillier encryption	1. Consumption of memory space and time complexity is very high
4	Ring signature	CryptoNote Monero	Scalability is poor Not feasible when the participants are increasing
5	Multipart computation	Millionaire problem	High computation costs between the parties involved Computation overhead and not suitable for large scale applications

Above mentioned mechanisms achieve privacy in blockchain network [7]. Moreover, these mechanisms process certain limitations which affects the growth rate of blockchain technology [8]. One major limitation is the use of third-party node in mixing services which leads to Denial of Service (DoS) attacks [9]. Performance and efficiency of existing mechanisms need to be improved. In addition, almost all existing mechanisms affects with high computational and storage overhead. So, building a large-scale application with blockchain is quite difficult with these issues. Hence, we propose a three layered protection scheme by integrating random address generation, content erasure mechanism and IntelSGX to overcome the limitations and to develop a highly efficient and secured privacy preserved blockchain network.

Implementing of the proposed privacy preserving scheme may bring drastic improvement in terms of efficiency, performance and security [17].

5   Proposed Method for Privacy Enhancement

Here we propose a three-layered protection scheme for privacy enhancement on blockchain network. The below mentioned figure represents the architecture of our proposed three-layered protection scheme (Fig. 4).

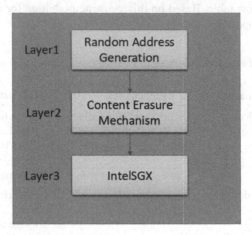

Fig. 4. Three layered protection architecture

5.1   Layer 1: Random Address Generation

This is the first layer of the proposed scheme. Whenever a transaction Tx is requested on blockchain network. Sender node initiates the transaction Tx and sends the request to the receiver node and the recipient node accepts the transaction [12]. Many privacy mechanisms are used in between to make the user anonymized. Here in our scheme both sender and receiver which is already assigned with an anonymous address is again randomized [13]. Every time when the sender node initiates the transaction Tx, Ty and so on random address is assigned to both the sender node and receiver node. For every transaction an unique address is assigned and this is an onetime unique address assigned for that particular transaction. This technique gives a complete anonymity to the sender and receiver nodes.

5.2   Layer 2: Content Erasure Mechanism

Transacting nodes are assigned with random address and the transaction are initiated. Transaction is carried out between the sender and the receiver. Once the transaction Tx is completed all the sensitive information stored on-chain are moved to off-chain storage [24]. Content erasure mechanism are used by which only hash pointers are made available on-chain. All the sensitive information related to the current transaction are moved to sidechains or off-chain storage. Other participants in the network will have access only to hash pointer where detailed information is kept hidden.

5.3   Layer 3: IntelSGX

Transaction information are moved from distributed ledger to the off-chain storage system. In off-chain storage these data need to be secured for which IntelSGX is used in third layer. Therefore, sensitive transaction data moved to sidechains or off-chain are secured using software guard extension (IntelSGX). SGX can be used to store the sensitive data by creating a secure storage space called enclave [11]. Only authorized users have access to enclave which is completely secured and free from attacker node [15]. Transaction information can be segregated into sensitive and non-sensitive information then sensitive information is moved to the enclaves of IntelSGX. It increases the privacy and security of the data by delivering more secured space that restricts disclosure of sensitive information. Integrating SGX improves security of transaction processing, consensus and smart contracts in blockchain technology.

Integration of this three-layered protection scheme improves the privacy on blockchain to a greater extent. In addition, we use privacy preserving proof of stake consensus mechanism for validation of transactions [27]. Inclusion of this consensus mechanism improves the validation time without data exploration. By this way privacy preserving mechanisms are included at different levels of a blockchain transaction. Complete user anonymity and highly efficient data privacy can be achieved.

6   Conclusion

Blockchain is a highly anticipated technology that are implemented to build many applications such as electronic voting, supply chain and identity preservation. Security and privacy issues in the blockchain restricts the growth rate of this technology to a greater extent [18]. In this paper, we analyzed the existing security and privacy challenges of blockchain technology. Major issues on the existing mechanisms are addressed. In addition, a three-layered protection scheme is also proposed for developing a highly efficient privacy preserving mechanisms. Existing privacy mechanisms of blockchain are prone to several limitations such us computation overhead, storage overhead and performance degradation [19]. Building up three layered protections may solve the existing limitations but the implementation feasibility of the proposed scheme need to be analyzed. Implementation and performance of the proposed scheme in large scale applications is the other major direction for the researchers to address.

References

1. Ron, D., Shamir, A.: Quantitative analysis of the full bitcoin transaction graph. In: Sadeghi, A.-R. (ed.) FC 2013. LNCS, vol. 7859, pp. 6–24. Springer, Heidelberg (2013). https://doi.org/10.1007/978-3-642-39884-1_2
2. Valenta, L., Rowan, B.: Blindcoin: blinded, accountable mixes for bitcoin. In: Brenner, M., Christin, N., Johnson, B., Rohloff, K. (eds.) FC 2015. LNCS, vol. 8976, pp. 112–126. Springer, Heidelberg (2015). https://doi.org/10.1007/978-3-662-48051-9_9
3. Wang, Z., Liu, J., Zhang, Z., Yu, H.: 'Full anonymous blockchain based on aggregate signature and confidential transaction.' J. Comput. Res. Develop. 55(10), 2185–2198 (2018). https://doi.org/10.7544/issn1000-1239.2018.20180430

4. Valenta, L., Rowan, B.: Blindcoin: blinded, accountable mixes for bitcoin. In: Brenner, M., Christin, N., Johnson, B., Rohloff, K. (eds.) Financial Cryptography and Data Security. FC 2015. LNCS, vol. 8976, pp. 112–126.Springer, Berlin (2015). https://doi.org/10.1007/978-3-662-48051-9_9
5. Sasson, E.B., et al.: Zerocash: decentralized anonymous payments from bitcoin. In: IEEE Symposium on Security and Privacy, pp. 459–474 (2014)
6. Dash is digital cash. https://www.dash.org//
7. Dong, G., Chen, Y., Fan, J., Hao, Y., Li, F.: 'Research on privacy protection strategies in blockchain application.' Comput. Sci. 46(5), 29–35 (2019). https://doi.org/10.11896/j.issn.1002-137X.2019.05.004
8. Liu, Z., Wang, D., Wang, B.: Privacy preserving technology in blockchain. Comput. Eng. Des. 40(6), 1567–1573 (2019). https://doi.org/10.16208/j.issn1000-7024.2019.06.012
9. Heilman, E., Alshenibr, L., Baldimisti, F., Scafuro, A., Goldberg, S.: Tumblebit: anuntrusted bitcoin compatible anonymous payment hub. In: Proceedings of the NDSS, pp. 1–15 (2017)
10. Goldwasser, S., Micali, S., Rackoff, C.: The knowledge complexity of interactiveproof systems. SIAM J. Comput. 18(1), 186–208 (1989). https://doi.org/10.1137/0218012
11. Zhang, H.G.: Research and development of trusted computing in China. In: ProceedingsAsia Pacific Trusted Infrastructure Technology Conference (APTC). IEEE Computer Society, New York, pp. 1–13 (2008)
12. Rajput, U., Abbas, F., Hussain, R., Eun, H., Oh, H.: A simple yet efficient approach to combat transaction malleability in bitcoin. In: Proceedings of the 9th International Workshop on Information Security Application, pp. 27–37. Springer, Cham (2015)
13. Van Saberhagen, N.: Cryptonote v2.0 (2013). https://static.coinpaprika.com/storage/cdn/whitepapers/1611.pdf
14. Chaum, D.L.: Untraceable electronic mail, return addresses, and digital pseudonyms. Commun. ACM 24(2), 84–90 (1981). https://doi.org/10.1145/358549.358563
15. Zhenyu, N., Fengwei, Z., Weisong, S.: A study of using TEE on edge computing. J. Comput. Res. Dev. 56(7), 1441–1453(2019)
16. Li, X., Niu, Y., Wei, L.., Zhang, C., Yu, N.: Overview on privacy protection in bitcoin. J. Cryptol. Res. 6(2), 133–149 (2019). https://doi.org/10.13868/j.cnki.jcr.000290
17. Bonneau, J., Narayanan, A., Miller, A., Clark, J., Kroll, J.A., Felten, E.W.: Mixcoin: anonymity for bitcoin with accountable mixes. In: Christin, N., Safavi-Naini, R. (eds.) FC 2014. LNCS, vol. 8437, pp. 486–504. Springer, Heidelberg (2014). https://doi.org/10.1007/978-3-662-45472-5_31
18. Sivaganesan, D.: Smart contract based industrial data preservation on block chain. J. Ubiquitous Comput. Commun. Technol. (UCCT) 2(01), 39–47 (2020)
19. Wang, H.: IoT based clinical sensor data management and transfer using blockchain technology. J. ISMAC 2(03), 154–159 (2020)
20. Bonneau, J., Narayanan, A., Miller, A., Clark, J., Kroll, J.A., Felten, E.W.: Mixcoin: anonymity for bitcoin with accountable mixes. In: 14th International Conference on Financial Cryptography and Data Security, LNCS, pp. 486–504 (2014)
21. Miers, C., Garman, M.G., Rubin, A.D.: Zerocoin: anonymous distributed e-cash from bitcoin. In: IEEE Symposium on Security and Privacy, pp. 397–411 (2013)
22. Fleder, M., Kester, M.S., Pillai, S.: Bitcoin transaction graph analysis (2015). https://arxiv.org/abs/1502.01657
23. Ruffing, T., Monero–Sanchez, P., Kate, A.: CoinShuffle: practical decentralized coin mixing for bitcoin. In: European Symposium on Research in Computer Security, pp. 345–364 (2014)
24. Wang, J., et al.: 'Analysis and research on SGX technology.' J. Softw. 29(9), 2778–2798 (2018). https://doi.org/10.13328/j.cnki.jos.005594
25. Todd. P.:Stealth Addresses. https://lists.linuxfoundation.org/pipermail/bitcoin-dev2014-January/004020.html. Accessed 6 Jan 2014

26. Hu, S., Cai, C., Wang, Q., Wang, C., Luo, X., Ren, K.: Searching an encrypted cloud meets blockchain: a decentralized, reliable and fair realization. In: Proceedings of the IEEE Conference on Computer Communications. (INFOCOM), Honolulu, pp. 792–800, April 2018. https://doi.org/10.1109/info-com.2018.8485890

27. Rivest, R.L., Shamir, A., Tauman, Y.: How to leak a secret. In: Proceedings of the 7th International Conference on the Theory and Application of Cryptology and Information Security, pp. 552– 565 (2001)

28. Chaum, D., Eugène, V.H.: Group signatures. In: Advances in Cryptology, pp. 257–265. Springer, Berlin (1991)

29. Rivest, R.L., Adleman, L., Dertouzos, M.L.: On data banks and privacy homomorphisms. Found. Secure Comput. 4(11), 169–180 (1978)

30. Tso, R., Liu, Z.-Y., Hsiao, J.-H.: Distributed E-voting and E- bidding systems based on smart contract.'Electronics 8(4), 422 (2019). https://doi.org/10.3390/electronics8040422

31. Wang, T.: A review of the study of secure multi-party computation. Cyberspace Secur. 5(5), 41–44 (2014)

A Comparative Analysis of Blockchain Platform: Issues and Recommendations-Certificate Verification System

K. Kumutha[1(✉)] and S. Jayalakshmi[2]

[1] Tagore College of Arts and Science, Chennai, India
[2] Department of Computer Applications, VISTAS, Chennai, India
jai.scs@velsuniv.ac.in

Abstract. The requirement for the blockchain era maintains growing, and many of the improvement platforms are going vital flow. Amongst those, businesses are further than ever keen to go for blockchain solutions, and they are inclined to the role of a large number of assets for that. There are many more blockchain development platforms are available. Ethereum and Hyperledger platform are greater famous today. The purpose of this research is to examine Ethereum and Hyperledger fabric theoretically and then to observe the problems and recommendations. This research suggests Hyperledger fabric Framework proposes a certificate verification system to avoid the fake and provide high level of security.

Keywords: Blockchain · Ethereum · Hyperledger fabric · Comparative analysis and certificate verification

1 Introduction

The future internet is blockchain technology. The first blockchain is Bitcoin which is introduced by Satoshi Nakamoto; it is Bitcoin that came into existence in 2009 [1]. Now the Bitcoin became more popular. Bitcoin is the most popular digital money used on peer to peer network in the case of the blockchain. Blockchain technology has the abilities that are Decentralized, Distributed, Secure and Faster, Transparent, and non-modifiable. These are more beneficial than the existing technologies. It is a linked list like data structure that maintains details of data and its transactions via a peer to peer network publically. The great advantage of blockchain enables most of the authors made to implement the educations system. It can store student details such as degree certificates and history of the provider and the address of the student's data in the network. The blockchain technology uses several consensus algorithms and common procedure execution through the distributed public ledger, business logic (smart contract), and cross-chain concepts. There are several blockchain development platforms are available. Each one has its own feature. This research paper analysis overall feature of Ethereum and Hyperledger fabric platform and based on this suggest suited platform implement the use

© IFIP International Federation for Information Processing 2021
Published by Springer Nature Switzerland AG 2021
V. Krishnamurthy et al. (Eds.): ICCIDS 2021, IFIP AICT 611, pp. 210–219, 2021.
https://doi.org/10.1007/978-3-030-92600-7_20

case. These techniques maintain a change of the data integrity by keeping the attributes of transactions such as time, space, and instantaneous multifunctional overlays with constraints such as recordable, traceable and determinable, cost and tradable procedure, etc.

This article has discussed the concepts of blockchain structure and its function and comparative analysis of blockchain platforms Ethereum and Hyperledger fabric. And also this paper presents the proposed certificate verification system design and its process.

2   Blockchain

A blockchain is a type of linked list data structure that is used in many distributed ledger technologies. It bundles all the changes made to the ledger (transactions) into packages called blocks and chains them together, using cryptographic hashing, providing an immutable record of all transactions from the genesis (first) block. The structure is presented in Fig 1.

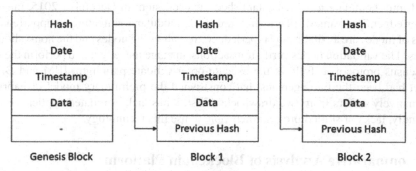

Fig. 1. Structure of a block

Each movement of records is secured with hashing SHA-256 algorithm and then all the transaction summary will be grouped and saved as blocks of data. Then the blocks are joined with the hash price of the previous block and so on and secured from tamper-proof. However, the hash of the block has calculated the use of all the transactions and the hash of a previous block. So, the hash of the ultimate block represents the total blockchain as one value, which makes comparing two chains really easy. As described, blockchain is an immutable report of all transactions. Consensus algorithm: The algorithm used for all the events to agree on a single world state is referred to as a consensus algorithm. Ethereum use the Proof of Work algorithm to reach consensus. All transactions in the previous blocks are validated and then, a new block is added to the blockchain. A consensus protocol consists of agreement, weak validity, robust validity and termination.

3   Blockchain Platform

The demand for the platform for blockchain creation is growing day by day in support of business have started to explore the blockchain based use cases. Another explanation why the number of blockchain platforms is growing is also the growth of dapp production. This article will assist in choosing the best blockchain application for project if who are new to this technology.

3.1 Ethereum

Ethereum is a global and open-source distributed and decentralized platform for all the decentralized applications. Ethereum is energetic since 2015, Vitali k Buterin is the co-founder. It is a permissionless blockchain platform. It has an inner cryptocurrency which is regarded as Ether. It additionally supports smart contracts. A smart contract is a programming code that is completed when an event is triggered. Ethereum operates these steps Block Validation, Node Discovery, Transaction Creation, and Mining in case of the transaction flow.

3.2 Hyperledger Fabric

The Linux Foundation, which established the ecosystem in December 2015, manages Hyperledger. The framework is open source and a modular architecture is supported. Two types of modes are available on Hyperledger: the validating nodes and the non-validating nodes. The validation nodes verify transactions, manage the ledger and perform the BFT consensus protocol. It follows the execute-order-validate paradigm. IBM and Digital Asset had been the two agencies that constructed the preliminary model of Fabric. It alternatively suffers from two drawbacks. First, it has lacks validated applications and secondly, lacks of skilled programmers able to use this technology.

4   Comparative Analysis of Blockchain Platform

This section compares Ethereum and Hyperledger Fabric platforms and table- 1 shows the overall features to gather and highlight the differences between them. And also find out the problems and recommendations of the blockchain development platform to implement the use case.

4.1 Architecture

Ethereum is the permissionless Blockchain for any kind of application. The transactions; are executed by using digital wallet "Gas" by the nodes and fully transparent. In contrast, Fabric affords bendy solutions for personal permissioned blockchain that permit safety and confidentiality. The latter is brought to the Fabric the use of channels that furnish unbiased ledgers on hand only to its users. So, it is viable to create multiple channels and join only some of the users to them. In this case, the ledger is private (cannot be accessed with the aid of non-registered users) and it is viable to share confidential records barring all the network noticing it.

4.2 Consensus Algorithm

Currently, Ethereum is designed to validate the transactions by using proof-of-work consensus algorithm, the place all the nodes agree upon a frequent reality and thereby the ledger. In contrast, Fabric lets in nodes to pick out between no-op (no consensus needed) and an agreement protocol (PBFT) whereby two or greater parties can agree on a key in such a way that both have an impact on the outcome. Thus, Fabric has fine-grained management over consensus and restrained get right of entry to transactions which consequence in extended overall performance scalability and privacy [2].

4.3 Applications

As Ethereum is public and totally obvious it ought to be efficaciously used for most of the ownership storages such as real estate and crypto-currency. Currently, it is feasible to use Ethereum as a fee technique in many methods such as grocery stores, ice-cream shops, online stores, etc. Fabric, on the different hand, gives a way to keep exclusive information that is required for any furnish chain. It permits many applications to be built on the pinnacle of a non-public blockchain, for example, educational certificate, electronic health records, or insurance, where information cannot be shared across the community however it needs to be accessible for precise contributors considering that it is totally private.

4.4 Smart Contract Language

One of the largest differences in blockchain platform is programming language. Ethereum uses Solidity language for writing smart contracts for applications and in contrast, Hyperledger uses more languages such as Go, JavaScript, Python and node.js etc. This ensures many programmers to write chaincode in Hyperledger instead of gaining knowledge of a new language [3]. Language help is certainly one of the biggest discrepancies. On the other hand, Ethereum supports languages that are explicitly designed to be used for writing Ethereum smart contracts, such as Solidity and Hyperledger Fabric, supporting more than one common programming language such as Go, JavaScript, and node.js. This takes us to a scenario where writing chain code is feasible.

4.5 Scalability

In Hyperledger fabric the transaction goes with the flow is divided into orderers, endorsers, and peers. Peers are the recipients of ordered sets of transactions, which they then commit to the network. Endorsers test the cryptographic signatures of a transaction to confirm that is steady with the state of the ledger [4]. Such a permission and modular method advantage the typical scalability of the network, which is the upper bound of transactions that can be processed in a given time-frame. This is special from Ethereum, where the roles played with the aid of the nodes that partake in the consensus protocol are identical, at the cost of its transaction throughput.

4.6 Smart Contracts and Chaincode

A "smart contract" is actually a program that runs on the Ethereum blockchain. In Ethereum, the User money owed can engage with a smart contract through submitting transactions that execute a feature described on the clever contract. Smart contracts can define rules, as a normal contract, and robotically enforce them by way of the code [4]. In Hyperledger, clever contracts are referred to as chaincode and, not like Ethereum smart contracts, don't require costs to be executed. Chaincode is developed with present programming languages, whereas in Ethereum clever contracts are developed with newly designed domain-specific languages such as solidity and accomplished with the aid of EVM. In the case of Hyperledger fabric, the well-supported language for the development of chaincode is Golang. In Fabric interoperability with Ethereum smart contracts is made possible.

5 Problem and Recommendations

The comparative analysis of blockchain structures indicates most of the applications are developed in Ethereum based; only a few papers spoke about the Hyperledger cloth due to lake of experts. And additionally there is a scalability issue in Ethereum compare to Hyperledger (Table 1). Legacy solutions Institutions have invested huge amounts of money into developing the infrastructure and placing up this software, integrating it into their processes. The team of workers has been skilled to operate these structures and it ought to be tough for them to transition to decentralized solutions.

5.1 Problems

Block chain's scalability: The greater information blocks are added to the blockchain, the slower it works, as it begins taking extra time to analyze all the facts history. Of course, distributed ledger applied sciences are becoming greater scalable nowadays, inventing new approaches of growing the speed of transaction processing [5, 10]. High preliminary costs: The fee of altering the environment, putting up the infrastructure, the value of blockchain development itself, as well as blockchain training for the staff. It could be pretty hard to foresee the true economic advantages of blockchain integration in the long run, which encompass reduced costs in management, record-keeping, and extra gains from introducing crypto repayments and rewards.

5.2 Recommendations

The order-execute architecture of the Ethereum blockchain framework slowdowns the transaction processing time. Smart contracts are packages going for walks on an Ethereum blockchain. They are immutable to change, and hence cannot be patched for bugs once deployed as public. Thus it is essential to ensure that the designed smart contracts are bug-free and well-developed before deployment. Change in smart contracts shows security, architecture, and/or usability problems. On the different hand, In Hyperledger Fabric supports non-public blockchain to use off-line CouchDB and on-line ledger. Changeable statistics are maintained in CouchDB and immutable facts like certificates on line ledger according to that maintain chaincode. Since its permission, architecture, scalability function insists to pick this platform to use the industrial applications [6]. Hyperledger Fabric is quickly gaining recognition and popularity. Most builders prefer it as it approves software enhancement and encourages writing, but it offers centralization and contains a membership carrier node that needs the member's identity, which is largely based on Proof of Work (PoW) and Proof of Stake (PoS) and is just a public blockchain.

Table 1. Comparative analysis of ethereum vs. hyperledger

Characteristics	Ethereum	Hyperledger
Type of blockchain	Public/Private	Private
Governance	Ethereum developers	Linux Foundation
Application type	It is general purpose so suited for B2C transactions	It is permissioned type hence suited for B2B transactions
Coin/token	Ether-ETH as a coin	No such a token/coin
Consensus protocol	PoW/PoS	No predefined consensus protocol, pluggable
Smart contract	Smart contract	Chaincode
Language	Solidity	Go, Java and Node.js
Nature of Transactions	Publically distributed anyone can access the ledger of transactions	Not public hence the authorized node can access the ledger
Partners	IC3, Microsoft, Accenture, Consens ys, Intel, Santader, CME Group, J.P. Morgan etc	Air bus, Accenture, American Express, Cisco, Daimier, J.P. Morgan, Intel, IBM, SAP etc
Throughput	Up to 20 Transactions per second (tps)	>2000 Transactions per second (tps)
Block-release timing	12 s	Configurable
Transaction size	Actual max size: 89 KB configurable	Maximum size configurable
Transaction rate	10 transactions/second	1000 transactions/second
Mining	PoW/PoS	N/A

Hyperledger comes with different structures such as Fabric, Iroha, Indy, Sawtooth and Burrow [12]. Though, Hyperledger Fabric (HLF) has been preferred to be utilized in this work due to its inherent privateness and role-based get right of entry to mechanism for getting access to the files that would be suiting to endorse the certificate verification system to prevent fake.

The architecture of the Hyperledger Fabric consists of peer nodes, nodes for ordering, and client applications. Two functions can be supported by a peer node: a committer when it holds the ledger by committing transactions and an endorser when it endorses the result after using the chaincode to simulate the transactions completed. In addition, a peer can be an undertaker for particular types of transactions when serving as an

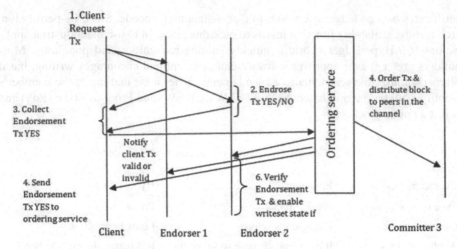

Fig. 2. Transaction flow in Hyperledger Fabric

endorser for others. Prior to committing it to the ledger, the ordering nodes take care of the order of the transactions in a block. This role can be centralized as effectively as it is decentralized. The function of peer and ordering nodes is comparable to the work performed in Ethereum by miners.

The feature of the transaction flows (Fig. 2) offers aspects of centralized as well as decentralized in Hyperledger Fabric blockchain science that continues facts with tamper-proof which will keep away from the intruder to get right of entry to and fakes the certificates from the blockchain. No one can access and adjust the certificates different than one who has got right of entry. So this effective function of the Hyperledger Fabric platform suggests that to put in force this blockchain technology to verify the diploma certificates and student's details insecure. The blockchain establishes a set of consensus and frequent operation mechanisms via the general ledger, smart contract, and cross-chain technology. The mechanism solidifies the data movement shaped by using time, space, and instant multidimensional overlays by way of programming to form recordable, traceable, determinable, priced, and tradable technological know-how constraints [2].

6 Proposed Work

The awards and degree certificates of the education institutes can include only the name of the group and the data of the recipient. There is a shortage of excellent anti-forgery mechanisms in this case, and as a result, several times the graduation certificates to be cast are often found. The architecture that would be suggested uses the permit to address the issue of false certificates is the Hyperledger Distributed Ledger Technology (DLT) to grant the following advantages [7, 8]. The blockchain network will issue the certificate in digital form through a distributed ledger, approved access, uniquely identifiable digital certificate and prevents forgery. In addition, it is very difficult to tamper with or manipulate the immutability nature of blockchain allows digital certificate in the distributed ledger and it is very straightforward to check the originality of digital

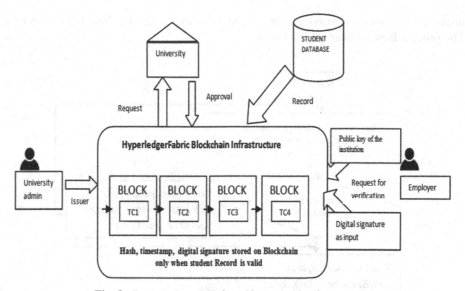

Fig. 3. Proposed model of certificate verification system

certificate. This system utilizes quite a number of features of blockchain science is a machine for industry-institute interaction the use of Hyperledger (Fig. 3).

Certificates furnished by education institutes, or certification units, universities, will have to be admitted to the scheme, and will be able to search through the system database. The authorities can provide a certificate through the system if students meet certain criteria. After the students have earned their certificates, they will be able to enquire about any certificates they have gained. The service provider is in charge of system renovation [13].

6.1 Process

To process the digital certificate verification system follows the following steps. The first step the blockchain must endorse the users. In this use case universities, institutes, students, and employers are users who verified by using multiple authentication systems through user id, password, biometric (face scanning, retina, fingerprint), and OTP generation. In step two, the valid user can upload the certificate details into the blockchain network with required certificate details and each created certificate will be stored in CouchDB which in turn will return the unique hash generated using the SHA-2S6 algorithm [14]. CouchDB used to store scanned certificate since only the essential details student id, serial no, date and time of issuing the certificate, issuing authority id, qualification along with the hash value. Once the block is created and then verified by a suitable consensus algorithm and the valid block is added to the blockchain network. Then a QR code, OTP, and query string will be created by the device to be affixed to a hard copy certificate to authenticate a hard copy of the certificate via the phone and website. The framework not only provides verification of the certificate, but also stores the certificate in digital form forever, provided the immutability of the distributed ledger [8]. And changing this certificate or producing a fake certificate with the same data is almost

218   K. Kumutha and S. Jayalakshmi

impossible. Thus, this proposed system can solve the issue of counterfeit certificates. The process flow is illustrated in Fig. 4.

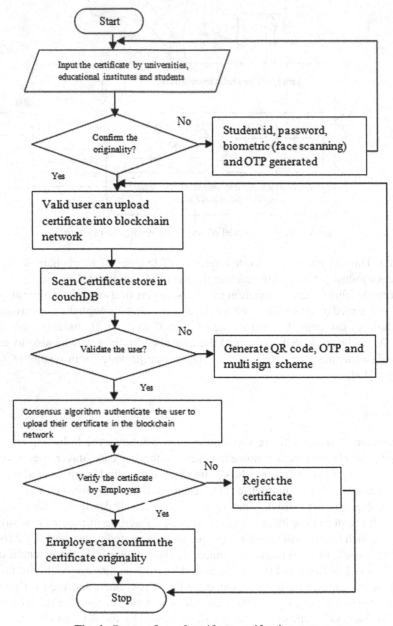

Fig. 4. Process flow of certificate verification system

7 Conclusion

The intention of this research is to take a look at Ethereum and Hyperledger blockchain platform features theoretically and then identified that the Hyperledger Fabric Platform is a perfect blockchain-based framework for verifying educational certificates. that specializes in specific issues is proposed [15]. Hyperledger Fabric, however, is designed for private use situations consisting of educational certificates verification inside the blockchain where every node participant have records handiest relevant for them. The risk of certificate forgery is minimized by using the proposed blockchain-based application. In this proposed system, the certificate application process and the automated certificate awarding process are open and transparent. Thus, businesses or organizations may ask for information from the blockchain network on any certificate. Thus, this proposed system guarantees the consistency and security of data. For future work, put in force the present day variations of these platforms and experimentally compare them the usage of distinct overall performance metric such as latency, throughput and transaction rate.

References

1. Nakamoto, S.: Bitcoin: A Peer-to-Peer Electronic Cash System, p. 8 (2008). http://Www.Bit coin.Org
2. Valenta, M., Sandner, P.: Comparison of Ethereum, Hyperledger Fabric and Corda (2017)
3. Macrinici, D., Cartofeanu, C., Gao, S.: Smart contract applications within blockchain technology: a systematic mapping study. Telemat. Inform. J. **35**, 2337–2354. https://doi.org/10.1016/j.tele.2018.10.004
4. Jiin-Chiou, N.-Y.L., Chi, C., Chen, Y.-H.: Blockchain and smart contract for digital certificate. In: Proceedings of IEEE International Conference on Applied System Innovation (2017)
5. https://www.blockcerts.org
6. Dinesh Kumar, K., Komathy, K., Manoj Kumar, D.S.: Blockchain technologies in financial sectors and industries. Int. J. Sci. Technol. Res. **8**(11), 942–946 (2019)
7. He, B.: An Empirical Study of Online Shopping Using Blockchain Technology. Department of Distribution Management, Takming University
8. Qiu, Z.: Digital certificate for a painting based on blockchain technology. Department of Information and Finance Management, National Taipei University of Technology, Taiwan (2017)
9. Diffie, W., Van Oorschot, P.C., Wiener, M.J.: Authentication and authenticated key exchanges. Des. Codes Crypt. **2**(2), 107–125 (1992)
10. "Ethereum project. https://github.com/ethereum/wiki/wiki/White-Paper. Accessed 13 Jan 2018
11. MIT Media Lab: What we learned from designing an academic certificates system on the blockchain. Medium, no. December (2016)
12. Androulaki, E., et al.: Hyperledger fabric: a distributed operating system for permissioned blockchain. In: Proceedings of the Thirteenth EuroSys Conference (2018)
13. https://www.indiatoday.in/education-today/featurephilia/story/how-students-and-employ ers-can-spot-and-eliminate-fake-degrees-1725931-2020-09-27
14. Dinesh Kumar, K., Senthil, P., Manoj Kumar, D.S.: Educational certificate verification system using blockchain. Int. J. Sci. Ttechnol. Res. **9**(03), (2020). ISSN: 2277-8616 82 ijstr©2020
15. Jerril Gilda, S.: Maanav Mehrotra-Blockchain for student data privacy and consent. In: 2018 International Conference on Computer Communication and Informatics (2018)

Author Index

Printed in the United States
by Baker & Taylor Publisher Services

Printed in the United States
by Baker & Taylor Publisher Services